Cognitive Technologies

The series Cognitive Technologies encompasses artificial intelligence and its subfields and related areas, such as natural-language processing and technologies, high-level computer vision, cognitive robotics, automated reasoning, multiagent systems, symbolic learning theories and practice, knowledge representation and the semantic web, and intelligent tutoring systems and AI and education. Cognitive science, including human and animal cognition and artificial life, is within the scope of this series, as is the integration of symbolic and subsymbolic computation.

The series includes textbooks, monographs, coherent thematic state-of-the-art collections, and multiauthor anthologies.

More information about this series at http://www.springer.com/series/5216

Wolfgang Hofkirchner • Hans-Jörg Kreowski
Editors

Transhumanism: The Proper Guide to a Posthuman Condition or a Dangerous Idea?

Editors
Wolfgang Hofkirchner
The Institute for a Global Sustainable Information Society (GSIS)
Vienna, Austria

Institute of Visual Computing and Human-Centered Technology
TU Wien
Vienna, Austria

Hans-Jörg Kreowski
Department of Computer Science
University of Bremen
Bremen, Germany

ISSN 1611-2482 ISSN 2197-6635 (electronic)
Cognitive Technologies
ISBN 978-3-030-56548-0 ISBN 978-3-030-56546-6 (eBook)
https://doi.org/10.1007/978-3-030-56546-6

This Springer imprint is published by the registered company Springer Nature Switzerland AG.
The registered company address is: Gewerbestrasse 11, 6330 Cham, Switzerland

Preface

Transhumanism is a worldwide philosophical and futuristic movement aiming to enhance the intellectual and physical capabilities of human beings beyond their current limits. Having its roots in the 1920s and 1930s, it has gotten quite some drive and attention in the last three decades. This is shown in an exemplary way by a variety of recently published books on this topic like Nick Bostrom's *Superintelligence: Paths, Dangers, Strategies*, Yuval Noah Harari's *Homo Deus: A Brief History of Tomorrow,* Olle Häggström's *Here Be Dragons: Science, Technology and the Future of Humanity*, Ray Kurzweil's *The Singularity Is Near: When Humans Transcend Biology* and *How to Create a Mind*, Stefan Lorenz Sorgner's *Schöner neuer Mensch* and Toby Walsh's *It's Alive*, to list but a few. Transhumanists intend to employ already existing and future technologies such as artificial intelligence, robotics, cognitive science, information technology, nanotechnology, biotechnology and others as human enhancement technologies. One of the leading protagonists of transhumanism, Max More, defines it as follows:

> Transhumanism is both a reason-based philosophy and a cultural movement that affirms the possibility and desirability of fundamentally improving the human condition by means of science and technology. Transhumanists seek the continuation and acceleration of the evolution of intelligent life beyond its currently human form and human limitations by means of science and technology, guided by life-promoting principles and values. (https://www.metanexus.net/h-true-transhumanism)

In contrast, critics like Francis Fukuyama consider transhumanism as one of "the world's most dangerous ideas". He writes:

> Nobody knows what technological possibilities will emerge for human self-modification. But we can already see the stirrings of Promethean desires in how we prescribe drugs to alter the behavior and personalities of our children. The environmental movement has taught us humility and respect for the integrity of nonhuman nature. We need a similar humility concerning our human nature. If we do not develop it soon, we may unwittingly invite the transhumanists to deface humanity with their genetic bulldozers and psychotropic shopping malls. (https://www.au.dk/fukuyama/boger/essay).

In this volume, the pros and cons of transhumanism from various points of view are discussed including information sciences, philosophy, sociology and technology.

In the mythologies all over the world, one encounters the idea of supernatural strength, invulnerability, eternal youth, invisibility, invincibility and immortality. Some proponents of transhumanism dream of a future in which all this will come true. And there are leading experts in nano-, bio- and information technologies as well as in cognitive science who formulate quite similar aims and objectives: the obligatory victory over Alzheimer disease and Parkinson disease, cleansing of wounds, blood, the lung, brain enhancement, soldiers who fight without fear, managers who need no sleep to be able to work 24 hours a day, 7 days a week for their companies, magic hoods and much more (see, e.g., the NSF/DOC-Sponsored Report on *Converging Technologies for Improving Human Performance* edited in 2003 by Mihail C. Roco and William S. Bainbridge).

The bulk of the contributions in this volume originates from a transhumanism workshop that took place at the IS4SI 2017 Summit *Digitalisation for a Sustainable Society* in Gothenburg, Sweden. The workshop was organized as an activity of the *IS4SI Special Interest Group Emergent Systems, Information and Society* which is also a working group of the *Leibniz Society of Sciences to Berlin e.V.* and has recently become a unit of *the Institute for a Global Sustainable Information Society* residing in Vienna as well as of the *Forum Computer Scientists and IT Professionals for Peace and Social Responsibility (FIfF e.V.), Germany.* Syed Mustafa Ali, Christopher Coenen, Wolfgang Hofkirchner, Roman M. Krzanowski with Kamil Trombik, Tomáš Sigmund and Christian Stary accepted our invitation to write a chapter of this book. Moreover, we invited Robert Ranisch, Alexander Reymann with Roland Benedikter as well as Stefan Lorenz Sorgner as authors of the dossier *Transhumanismus und Militär* that appeared in 2018 as an addendum to the journal *Wissenschaft und Frieden.* To get an even broader view, Adriana Braga and Bob Logan, Olle Häggström, Klaus Kornwachs, Giuglielmo Papagni, Rainer Rehak and Britta Schinzel were invited to contribute in addition.

The volume is organized into four parts concerning philosophical, military, technological and sociological aspects of transhumanism. But the association of the chapters to those categories is not at all uniquely determined in every case. Moreover, other meaningful categories might fit as well.

We are grateful to the publisher and in particular to Ronan Nugent for his patient support of the editing process of this volume.

June 30, 2020
Vienna, Austria — Wolfgang Hofkirchner
Bremen, Germany — Hans-Jörg Kreowski

Contents

Contributors

Syed Mustafa Ali School of Computing and Communications, The Open University (UK), Milton Keynes, UK

Roland Benedikter Center for Advanced Studies, Eurac Research, Bozen-Bolzano-Bulsan, Italy
Multidisciplinary Political Analysis, Willy Brandt Centre, University of Wroclaw, Wroclaw, Poland

Adriana Braga Department of Social Communication, Pontifícia Universidade Católica do Rio de Janeiro (PUC-RJ), Rio de Janeiro, RJ, Brazil

Christopher Coenen Institute for Technology Assessment and Systems Analysis (KIT-ITAS), Karlsruhe Institute of Technology, Karlsruhe, Germany

Olle Häggström Chalmers University of Technology, Gothenburg, Sweden
Institute for Futures Studies, Stockholm, Sweden

Wolfgang Hofkirchner The Institute for a Global Sustainable Information Society (GSIS), Vienna, Austria
Institute of Visual Computing and Human-Centered Technology, TU Wien, Austria

Klaus Kornwachs Humboldt Center for Humanities, University of Ulm, Ulm, Germany

Roman M. Krzanowski The Pontifical University of John Paul II, Cracow, Poland

Robert K. Logan Department of Physics, University of Toronto, Toronto, ON, Canada

Guglielmo Papagni Institute of Management Science, TU Wien, Vienna, Austria

Robert Ranisch Research Unit "Ethics of Genome Editing", Institute for Ethics and History of Medicine, University of Tübingen, Tübingen, Germany
International Centre for Ethics in the Sciences and Humanities (IZEW), University of Tübingen, Tübingen, Germany

Rainer Rehak Weizenbaum Institute for the Networked Society, Berlin, Germany

Alexander Reymann Görgeshausen, Germany

Britta Schinzel Computer Science and Social Research, University of Freiburg, Freiburg, Germany

Tomáš Sigmund Prague University of Economics and Business, Prague, Czech Republic

Stefan Lorenz Sorgner John Cabot University, Rome, Italy

Christian Stary Business Informatics – Communications Engineering, Johannes Kepler University of Linz, Linz, Austria

Kamil Trombik The Pontifical University of John Paul II, Cracow, Poland

Part I
Philosophical Aspects

Chapter 1
Aspects of Mind Uploading

Olle Häggström

Abstract Mind uploading is the hypothetical future technology of transferring human minds to computer hardware using whole-brain emulation. After a brief review of the technological prospects for mind uploading, a range of philosophical and ethical aspects of the technology are reviewed. These include questions about whether uploads will have consciousness and whether uploading will preserve personal identity, as well as what impact on society a working uploading technology is likely to have and whether these impacts are desirable. The issue of whether we ought to move forward towards uploading technology remains as unclear as ever.

1 Introduction

According to transhumanism, the current form of *homo sapiens* should not be thought of as the end product of evolution, but rather as a transitionary state on the path towards posthuman life forms that we can achieve by enhancing ourselves, e.g. pharmacologically, genetically or by direct brain–computer interfaces. Transhumanists claim that such a development is not only possible but desirable. In this, they are opposed by so-called bioconservatives, who maintain that it is undesirable or even forbidden, pointing at a variety of reasons, religious as well as secular. See Bostrom and Savulescu (2009) and Hansell and Grassie (2011) for collections of papers representing both sides of the transhumanism vs. bioconservatism debate, or Chapter 3 of Häggström (2016) for my own attempt at a fair and balanced summary of the key issues.

The ultimate dream for many transhumanists is *uploading*, defined here as the transferring of our minds to computer hardware using *whole-brain emulation*, the idea being that a good enough simulation of a human brain simultaneously gives a

O. Häggström (✉)
Chalmers University of Technology, Gothenburg, Sweden

Institute for Futures Studies, Stockholm, Sweden
e-mail: olleh@chalmers.se

W. Hofkirchner, H.-J. Kreowski (eds.), *Transhumanism: The Proper Guide to a Posthuman Condition or a Dangerous Idea?*, Cognitive Technologies,
https://doi.org/10.1007/978-3-030-56546-6_1

simulation of the human mind, and that if the simulation is sufficiently detailed and accurate, then it goes from being a mere simulation to being in all relevant aspects an *exact replica* of the mind—an emulation. What precisely is the correct meaning of "all relevant aspects" is open to debate, and while this chapter will touch upon this issue, no pretension is made to settle it.

There are several reasons why we might want to upload. An uploaded mind will, at any given time, consist of a finite string of 0's and 1's, making long-distance travel as easy as the transfer of computer files. We will no longer be stuck with our fragile flesh-and-blood bodies, but can migrate between robot bodies at will, or into virtual worlds. Furthermore, uploading allows us to easily make backup copies of ourselves. A sufficient number of such backup copies of ourselves will not make us literally immortal (because, e.g., a civilization-scale catastrophe might destroy all of them), but it vastly improves the prospects for astronomically long lives, and will likely make us much less concerned about being killed in accidents or violent attacks, because all one loses from such an event is the memories acquired since the last time one made a backup copy. An even more mind-blowing variant of this is the idea of making copies not meant for idle storage but for being up and running in parallel with the original.

These are some of the ideas that make uploading an attractive idea to many (but not all) transhumanists and futurologists. There is, however, a variety of further concerns regarding whether uploading technology can be successfully developed—and if so, whether it *should* be done. The purpose of this chapter is to review what I consider to be the key issues in this discussion, drawing heavily (but far from exclusively) on Sections 3.8 and 3.9 of Häggström (2016).

I will begin in Sect. 2 with questions about what is needed for an *operationally successful* uploading technology, meaning that if I upload, the upload should correctly reproduce my behaviour: it should look to any outside observer as if the upload is actually me. Is it reasonable to expect that such a technology is forthcoming? If yes, when? Can we even hope for *non-destructive* uploading, meaning that the procedure leaves my brain (and body) intact, or would we need to settle for *destructive* uploading?

Then, in Sects. 3 and 4, which can jointly be viewed as the core of this chapter, I will discuss ways in which an operationally successful uploading technology might nevertheless be a *philosophical failure*, in the sense that although if I upload everything looks good from the outside, uploading nevertheless does not give *me* what I want. One way in which this might happen is if the upload fails to have consciousness (a possibility treated in Sect. 3). Another is that even if the upload has consciousness, it might fail to give me what I want by not being *me*, but merely someone else who shares my memories, my psychological traits and so on (Sect. 4).

Among the various ethical concerns involved in whether or not we should move forward towards an uploading technology, I collect some of them in Sect. 5, except for what is perhaps the biggest one, namely what we can expect a society with a widely established uploading technology to be like, and whether such a society is worth wanting. That issue is treated separately in Sect. 6, based heavily on Hanson's (2016a) recent book *Age of Em*, which offers a generous range of predictions about what to expect from a society of uploads, or of *ems* (short for emulated minds) as he calls them. Finally, in Sect. 7, I wrap up with some concluding remarks.

2 Is It Technically Doable?

Timelines for the emergence of human- or superhuman-level artificial general intelligence (AGI) are so uncertain that epistemically reasonable predictions need to smear out their probability distributions pretty much all over the future time axis; see, e.g., Bostrom (2014), Grace et al. (2017) and Yudkowsky (2017). In contrast, we know enough about how complicated and involved whole-brain emulation is to at least be able to say with confidence that a technology for mind uploading is not forthcoming in the next 5 or 10 years (the only scenario I can think of that might reasonably falsify this prediction is if we suddenly discover how to create super-intelligent AGI and this AGI decides to engage in developing mind uploading technology). While we might consider mind uploading to be the logical endpoint if we extrapolate two of today's largest and most prestigious ongoing research projects (the European Union's Human Brain Project and the White House BRAIN Initiative), neither of these has mind uploading among its explicit goals.

Kurzweil (2005) has done much to popularize the idea of mind uploading, and true to his faiblesse for unabashedly precise predictions of technological timelines, he states with confidence that uploading will have its breakthrough in the 2030s. This prediction is, however, based on a variety of highly uncertain assumptions. Sandberg and Bostrom (2008) are better at admitting these uncertainties, and their sober report still stands out as the main go-to place on the (future) technology of whole-brain emulation. They systematically explore a wide range of possible scenarios, and break down much of their analysis into what they consider the three main ingredients in whole-brain emulation, namely *scanning*, *translation* and *simulation*. Besides these, there is also the need to either give the emulation a robotic embodiment with suitable audiovisual and motoric input/output channels or embed it in some virtual reality environment, but this seems like a relatively easy task compared to the main three (Bostrom 2014). Briefly, for what is involved in these three, we have the following summaries adapted from Häggström (2016).

- *Scanning* is the high-resolution microscopy needed to detect and register all the relevant details of the brain. While huge uncertainty remains as to the level of resolution needed for this, it seems likely that the necessary level has already been attained, since techniques for seeing individual atoms are becoming increasingly routine. Still, the speed and parallelization likely needed to avoid astronomical scanning times is not there yet. Sandberg and Bostrom (2008) focus mainly on techniques that involve cutting up the brain in thin slices that are scanned separately using one or other of the microscopy technologies that may be available (magnetic resonance imaging, sub-diffraction optics, X-ray, electron microscopy and so on). This seems like a clear case of destructive scanning, so whatever uploading procedure it forms part of will obviously have to be destructive as well. Non-destructive scanning is even more challenging. A speculative suggestion, advocated by Kurzweil (2005) and building on the work of Freitas (1999), is to use nanobots that enter the brain in huge numbers, observe the microscopic structures in there and report back.

- *Translation* is the image analysis and other information processing needed to turn the scanned data into something that can be used as the initial state for the simulation. The amount needed seems to be huge; Sandberg and Bostrom (2008) list some of the stuff that needs to be done: "Cell membranes must be traced, synapses identified, neuron volumes segmented, distribution of synapses, organelles, cell types and other anatomical details (blood vessels, glia), identified".
- *Simulation* requires large amounts of hardware capability in terms of memory, bandwidth and CPU, presumably in a massively parallel architecture. Exactly how large these amounts need to be depends, of course, on the level of microscopic detail required for a satisfactory emulation. It may also depend on the amount of understanding we have of the higher-level workings of the brain; the less we have of that, the more critical is it to get the microscopic details exactly right, requiring more computer power. Whole-brain emulation is often envisioned as a pure bottom-up (and brute force) project with no need for such higher-level understanding, in which case it will seem to us more or less as a black box (although no more than our brain already does). A more nuanced view of future technologies should allow for the possibility of significant amounts of top-down design, based on advances in the neuroscientific understanding of how our minds work at various levels above the microscopic details.

Another possibility, interlacing the three main ingredients listed above, is the piece-by-piece replacement of the brain by electronic devices maintaining the same functionality and interfacing with the rest of the brain as the replaced tissue; this idea was pioneered by Moravec (1988). Eventually all that remains is electronics, and the information stored can then be moved at will. As far as the end result is concerned, this should probably count as destructive uploading, but some thinkers intuit the gradualness of the procedure as a less scary form of destruction and are more willing to believe in the survival of personal identity (whatever that means; see Sect. 4).

While alternatives are not obviously impossible, it seems likely that the first uploading method to become technically available will be destructive, and based on scanning brain slices. As to timelines, Bostrom (2014) sticks to the (tentative) verdict of Sandberg and Bostrom (2008), which he summarizes as "the prerequisite capabilities might be available around mid-century, though with a large uncertainty interval".

3 Consciousness

Consider the case of destructive uploading, in which case my brain is destroyed, and replaced by a whole-brain emulation on computer hardware. It might be that, no matter how well the technology discussed in the previous section is perfected, if I upload I will still not be around to experience the thoughts and the doings of the upload. It is hardly inconceivable that if my brain is frozen and cut up in slices, I simply die, no matter what is done to the information stored in my brain. This would make the prospect of uploading a lot less attractive than if I could look forward to a rich life, full of experiences, in my new existence as an upload. (Still, it might not *always* make it *entirely* unattractive, as for instance I might nevertheless choose to upload if I am about to die anyway of some incurable disease while still having some life projects that I would like to complete, such as proving the Riemann hypothesis or raising my children.)

One way in which destructive uploading would cut me off from future experiences is if the upload simply lacks consciousness. Perhaps, it is just not possible to harbour consciousness on a digital computer. Whether or not this is the case is an open question, along with the more general issues of what consciousness really is, and how it arises in a physical universe. Philosophers of mind cannot be expected to settle these matters anytime soon; see, e.g., Chalmers (1996), McGinn (2004), Searle (2004) and Dennett (2017) for some of the different and partly conflicting views held by leading contemporary philosophers. Until this is sorted out, any discussion of whether uploads will be conscious or not will have to be speculative to a considerable extent. But the topic is important, so we should not let this deter us—perhaps we can make progress!

My personal inclination, while far from being an unshakable belief, is towards accepting uploads as conscious beings. My reason is based on the following thought experiment (Häggström 2016). I judge most or all of my human friends to be conscious beings, including my friend Johan. Let us imagine him a few decades from now taking me by surprise by removing the top of his skull and demonstrating that it does not contain the expected jellyish object known as a brain, but is instead full of electronic computer hardware. Note (crucially for my argument) that if we accept my position, based on the work reviewed in Sect. 2, that mind uploading is in principle operationally feasible, then something like this is a fairly plausible scenario. Johan's display would not change my verdict that he is conscious, because my (current) conviction that he is conscious is based not on beliefs about his inner anatomy, but on his speech and behaviour, which will be as irresistibly conscious-seeming in the thought experiment as it is today. It seems to me that this is how, in general, we judge others to be conscious. If those reasons are right, then we should grant consciousness also to uploads behaving like us, such as Johan in the thought experiment.

I admit that this argument is nowhere near a demonstration that uploads will be conscious, and perhaps it should not even count as a good reason. Perhaps, it just means that our spontaneous intuitive reasons for judging others as conscious are no good. But if we stick to those reasons, then the consistent thing to do is to treat uploads the same (analogously to how already today we grant consciousness to, say, people whose gender or ethnicity differs from our own). To me, this line of thinking is further supported by a moral argument: we do not know who is conscious or who is not (heck, I do not even know that my friend Johan is conscious), but it seems to me that anyone who acts conscious is entitled to our benefit of the doubt. This is in line with Sandberg's (2014) Principle of Assuming the Most: "Assume that any emulated system could have the same mental properties as the original system and treat it accordingly". In contrast, Fukuyama (2002) simply denies that uploads can be conscious, and goes on to say (in a way that is clearly meant not only descriptively but also normatively) that "if we double-crossed [an upload] we would feel no guilt [...] and if circumstances forced us to kill him [...] we would feel no more regret than if we lost any other valuable asset, like a car or a teleporter".

The main idea to be found in the literature in favour of uploads being conscious is the so-called *computational theory of mind* (CTOM). Roughly, it states that what matters for consciousness is not the material substance itself but its organization ("it ain't the meat, it's the motion", as Sharvy 1985 put it), and that the organization that produces consciousness is the right kind of information processing; what precisely "the right kind of" means is typically left unspecified for the time being. Of the philosophers mentioned above, Dennett is a proponent of CTOM, Chalmers looks agnostically but fairly favourably upon it and Searle thinks (for reasons we will have a look at a few paragraphs down) it badly mistaken, while McGinn adheres to the so-called mysterian view that understanding the fundamentals of consciousness is not within reach of the human cognitive machinery in much the same way that a dog can never understand Fermat's last theorem. Fukuyama (2002) rejects CTOM for no other reason than the (unsupported and probably false) statement that "it works only by denying the existence of what you and I and everyone else understand consciousness to be (that is, subjective feelings)".

Arguments like the thought experiment about Johan above tilt me towards a favourable view of CTOM. But there are also arguments against it. Fairly typical is the argument by Pigliucci (2014) who, in his contribution to the collection by Blackford and Broderick (2014), repeatedly asserts that "consciousness is a biological phenomenon", and complains that Chalmers (2014) in the same volume "proceeds *as if* we had a decent theory of consciousness, and by that I mean a decent *neurobiological theory*" (emphasis in the original). Since CTOM is not a neurobiological theory, it does not pass Pigliucci's muster and must therefore be wrong.

Or so the argument goes. As first explained in Häggström (2016), I do not buy it. To expose the error in Pigliucci's argument, I need to spell it out a bit more explicitly than he does. Pigliucci knows of exactly one conscious entity, namely himself, and he has some reasons to conjecture that most other humans are conscious as well, and furthermore that in all these cases the consciousness resides in the brain (at least to a large extent). Hence, since brains are neurobiological objects, consciousness must be a (neuro-)biological phenomenon. This is how I read Pigliucci's argument. The problem with it is that brains have more in common than being neurobiological objects. For instance, they are also material objects, and they are computing devices. So rather than saying something like "brains are neurobiological objects, so a decent theory of consciousness is neurobiological", Pigliucci could equally well say "brains are material objects, hence panpsychism", or he could say "brains are computing devices, hence CTOM", or he might even admit the uncertain nature of his attributions of consciousness to others and say "the only case of consciousness I know of is my own, hence solipsism". So what is the right level of generality? Any serious discussion of the pros and cons of CTOM ought to start with the admission that this is an open question. By simply postulating from the outset what the right answer is to this question, Pigliucci short-circuits the discussion, and we see that his argument is not so much an argument as a naked claim.

There are somewhat better anti-CTOM arguments around than Pigliucci's. The most famous one is *the Chinese room argument* of Searle (1980). The following summary and discussion of the argument is mostly taken from Häggström (2016). The argument is a *reductio ad absurdum*: assuming correctness of CTOM, Searle deduces the in-principle possibility of a scenario that is so crazy that the assumption (CTOM) must be rejected. Here is how he reasons:

Assume CTOM. Then a computer program can be written that *really* thinks, that *really* understands and that *really* is conscious, as opposed to merely giving the outward appearance of doing so. Suppose, for concreteness, that the program speaks and understands Chinese. (Since there are humans who can do this, CTOM implies that there are such computer programs as well.) Let us now implement this program in a slightly unusual way—as opposed to on an electronic computer. Instead of a computer, there is a room containing John Searle himself (who does not understand Chinese), plus a number of large stacks of paper. The first stack gives precise instructions in English (corresponding to the computer program) for what Searle (corresponding to the CPU) should do, while the other stacks play the role of computer memory. The room furthermore has two windows, one where people outside can provide strings of Chinese symbols as input to the system, and another where Searle delivers other such strings as output—strings that he produces by following the step-by-step instructions in the first stack. The same information processing is going on here as would have been the case in the computer, so if CTOM is true, then Searle understands Chinese, which is crazy, because all he knows is how to mindlessly follow those step-by-step instructions, so CTOM must be wrong.

There are several ways in which a CTOM proponent might respond to this attempt at a *reductio*. Personally, I am inclined to side with Hofstadter and Dennett (1981) in holding forth what Searle calls *the systems reply*, which is to say that in the proposed scenario, it is not Searle who understands Chinese, but the whole system, consisting of the room, plus the stacks of paper, plus Searle. Searle is just one of the components of the system and does not understand Chinese any more than a single neuron can be found in my brain that understands the statement of Fermat's last theorem.

Searle (1980) does have a retort to the systems reply, namely that the thought experiment can be modified so as to get rid of the room. Instead, he himself memorizes the entire rulebook and carries out the step-by-step manipulations in his head. Again we have the same information processing going on, so CTOM implies that Searle understands Chinese, a consequence that the systems reply can no longer explain away, because this time the system consists of nothing but Searle. But still Searle does not understand Chinese—all he does is mindless step-by-step symbol manipulation—so CTOM must be wrong.

I recommend the discussion that followed in Hofstadter and Dennett (1981), Searle (1982a), Dennett (1982) and Searle (1982b) to those readers who enjoy a good fight, and as a fascinating example of the amount of heat and anger a philosophical discussion can generate. Hofstadter and Dennett suggest that Searle has failed to imagine the astounding amount of learning needed to "swallow up the description of another human being" and whether that can be done without thereby achieving an understanding of whatever it is that this other human being understands. With a particularly inflammatory phrase, they stress that "a key part of [Searle's] argument is in glossing over these questions of orders of magnitude". Searle (1982a, b) replies, but offers no amendment to this alleged glossing over, and no further clue as to exactly *why*, under these extreme circumstances, the individual will not understand Chinese. Maybe he will, maybe he would not.

When I try to imagine the outcome of this extreme thought experiment, the most natural interpretation that comes to mind is in terms of multiple personality disorder (Häggström 2016). To us outside observers, it will look like we are faced with two persons inhabiting the same body: English-speaking Searle who tells us (in English) that he does not understand a word of Chinese, and Chinese-speaking Searle who insists (in Chinese) that he does understand Chinese. Why in the world should we trust only the former and not the latter?

Bergström (2016) will have none of this, and says that in this situation, "Searle knows Chinese only in the same sense that I 'know' how to play grandmaster-level chess, namely if I have a grandmaster next to me telling me which moves to make" (my translation). Here, when speaking of "Searle", Bergström obviously refers to English-speaking Searle, whose perspective he is eager to entertain, while he entirely ignores that of Chinese-speaking Searle. This asymmetry strikes me as terribly arbitrary. It is almost as if he is taking for granted that Chinese-speaking Searle does not even *have* a perspective, a shocking violation of the benefit-of-the-doubt moral principle I suggested earlier in this section.

These examples from the literature only scratch the surface of what has been said about CTOM, but they serve to illustrate how little we know about the origins and role of consciousness in the physical universe. CTOM may come across as counterintuitive, something that the Chinese room argument does make some way towards showing, but also something that Schwitzgebel (2014) argues is bound to hold for *any* theory of consciousness—counterintuitive enough to warrant in his opinion the term *crazy*. And while CTOM may well even be wrong, it survives quite easily against currently available attempts to shoot it down, and it strikes me as sufficiently elegant and consistent to be a plausible candidate for a correct theory about how consciousness arises in the physical world. This is encouraging for those who hope for conscious uploads, but the jury is likely to remain out there for a long time still.

4 Personal Identity

Even if it turns out uploads are conscious, I might still hesitate to undergo destructive uploading, because if the upload is not *me*, but merely a very precise copy of me (my personality traits, my memories and so on), then destructive uploading implies that I die. So will the upload be me? This is the problem of personal identity. Throughout most of this section, I will (mostly for the sake of simplicity) discuss personal identity in the context of *teleportation* rather than uploading; the cases are very similar and the arguments can be easily transferred from one case to the other.

Teleportation is a hypothetical (and controversial) future means of transportation. An individual's body (including the brain) is scanned, and the information thus obtained is sent to the destination, where the body is reassembled—or, if you prefer to view it that way, a new identical body is assembled. (Whether this distinction is substantial or merely terminological is part of the problem of personal identity.) It is an open problem whether it will ever be doable, but we will assume for the sake of the argument that we have a teleportation technology good enough so that the reassembled individual, including his or her behaviour, is indistinguishable from the original. As with uploading, we distinguish between *destructive teleportation* (where the scanning procedure involves the destruction of the original body) and *non-destructive teleportation* (where the original body remains intact).

Teleportation has been commonplace in science fiction for more than half a century, with occasional appearances before that, such as in Richard Wagner's 1874 opera *Der Ring des Nibelungen*. Among pioneering philosophical treatments of the issue of whether personal identity survives destructive teleportation we find Lem (1957) and Parfit (1984). Resolving the issue empirically seems difficult (as anticipated already by Lem) and may perhaps even be impossible: even in a world where such a technology is widely established, it does not help to ask people who have undergone teleportation whether or not they are the same person as before they teleported, because no matter what the true answer is, they will all *feel* as if they are the same person as before, and respond accordingly (provided they are honest). Faced with these bleak prospects for resolving the issue empirically, one way out, in the logical positivist tradition, is to rule it out as ill-posed or meaningless. Let us still have a go at a philosophical analysis. The following formalism, tentatively distinguishing two kinds of survival, is taken from Häggström (2016).

Imagine that today is Tuesday, I am in Gothenburg, but need to attend a meeting in New York on Thursday. Tomorrow (Wednesday) is travel day, and I need to choose between (destructive) teleportation and old-fashioned air travel. The former is far more convenient, but if it means that I die then I prefer to go by air. Thus: if I take the teleporter, will I survive until Thursday? That may depend on the exact meaning of survival, and as a preliminary step towards sorting out this thorny issue, let us distinguish *σurvival* from *Σurvival*, as follows:

σurvival: I σurvive until Thursday if, on Thursday, there exists a person who has the same personality traits and memories and so on, as I have today, allowing for some wiggling corresponding to the amount of change we expect a person to go through over the course of a few days.

Σurvival: I Σurvive until Thursday if (a) I σurvive until Thursday, and (b) the situation satisfies whatever extra condition is needed for the guy in New York to really be me.

It is clear that I will σurvive the teleportation, but will I also Σurvive? That is unclear, due to the definition of Σurvival being incomplete, or one might even say fuzzy—what exactly *is* that condition in (b)? The fuzziness is intentional, because I honestly do not have any good idea for what that property might be. In my view, any proposal for a nonvacuous condition in (b) needs to be backed up by a good argument, because otherwise Occam's razor (Baker 2010) should compel us to simply drop the condition, and conclude that Σurvival coincides with σurvival—in which case teleportation is safe.

We may find a hint about what the condition might be in what is perhaps the most common objection to the idea of survival of personal identity under destructive teleportation, namely the comparison with non-destructive teleportation summarized by Yudkowsky (2008) as follows:

> Ah, but suppose an improved Scanner were invented, which scanned you non-destructively, but still transmitted the same information to Mars. Now, *clearly*, in this case, *you, the original* have simply stayed on Earth, and the person on Mars is *only a copy*. Therefore [teleportation without the improved scanner] is actually murder and birth, not *travel* at all—it destroys the original, and constructs a copy! [Italics in the original]

While Yudkowsky's purpose here (just like mine) is to state the argument in order to shoot it down, I do not think his eloquent summary of it (including his sarcastic "*clearly*") is at all unfair. Chalmers (2014) and Pigliucci (2014) are among the many writers who consider it; Chalmers attaches some weight to it but considers it inconclusive, while Pigliucci regards it as conclusive and claims that "if it is possible to do the transporting or uploading in a non-destructive manner, *obviously* we are talking about duplication, not preservation of identity" (italics in the original). Note the convenient ambiguity of Pigliucci's "*obviously*" (echoing Yudkowsky's "*clearly*"), and that the term is warranted if it refers to "duplication" but not at all so if it refers to "preservation of identity". If σurvival is all there is to survival, then there is no contradiction in surviving in two bodies.

Pigliucci fails to spell out his argument, but it seems safe to assume that he (along with other proponents of the destructive vs. non-destructive comparison argument) does think there is some nonvacuous condition in (b). Since a major asymmetry between the two bodies in the non-destructive teleportation case is that only one of them has a continuous space-time trajectory from the original (pre-teleportation) body, my best guess at what condition he implicitly refers to is CBTST, short for *continuity of my body's trajectory through space-time* (Häggström 2016). If CBTST is the property in (b), then clearly I will *not* Σurvive the teleportation to New York. But is CBTST really crucial to personal identity? Perhaps if preservation of personal identity over time requires that I consist of the same elementary particles as before, but this seems not to be supported by fundamental physics, according to which elementary particles do not have identities, so that claims like "these electrons are from the original body, while those over there are not" simply do not make sense (Bach 1988, 1997). Since our bodies consist of elementary particles, postulating the corresponding asymmetry between the two bodies coming out of the non-destructive teleportation procedure is just as nonsensical. And even if physics did admit elementary particles with individual identities, the idea of tying one's personal identity to a particular collection of elementary particles seems indefensible in view of the relentless flux of matter on higher levels. Says Kurzweil (2005):

> The specific set of particles that my body and brain comprise are in fact completely different from the atoms and molecules that I comprised only a short while ago. We know that most of our cells are turned over in a matter of weeks, and even our neurons, which persist as distinct cells for a relatively long time, nonetheless change all of their constituent molecules within a month. [. . .] The half-life of a microtubule is about ten minutes. [. . .]
>
> So I am a completely different set of stuff than I was a month ago, and all that persists is the pattern of organization of that stuff. The pattern changes also, but slowly and in a continuum. I am rather like the pattern that water makes in a stream as it rushes past the rocks in its path. The actual molecules of water change every millisecond, but the pattern persists for hours or even years.

Proponents of CBTST—or of any other nonempty condition—as being crucial to (b) need to spell out good arguments for why this would be the case. Until they do, the reasonable stance seems to be that survival is simply σurvival, in which case teleportation is safe. The view that survival is just σurvival may seem counterintuitive or even appalling—why should I care as much as I do about the person tomorrow who claims to be me if all that connects his personal identity to mine is mere σurvival? See Blackmore (2012) for a beautiful attempt to come to grips with this in our day-to-day existence; also Parfit (1984) arrives at a similar position.

In order to connect this teleportation discussion back to the issue of uploading, assume now that we accept that destructive teleportation is safe in the sense of preserving personal identity, and assume furthermore that after careful reflection of the issues discussed in Sect. 3 we also accept that uploads are conscious. Should we then conclude that we would survive destructive uploading? While the two assumptions do offer some support in this direction, the conclusion does not quite follow, because condition (b) for Σurvival might for instance be that (the physical manifestation of) the person on Thursday is made of the same kind of substance as the one on Tuesday. The desired conclusion does, however, seem to follow if we strengthen the survival-under-teleportation assumption to Σurvival = σurvival.

5 Ethical Concerns

With the development of unloading technology, we will encounter difficult ethical problems, and the corresponding legal ones. For instance, Bostrom (2014) notes that, just like with any other technology, "before we would get things to work perfectly, we would probably get things to work imperfectly". This might well amount to creating a person with (an upload analogue of) severe brain damage resulting in horrible suffering. Would it be morally permissible to do so? Metzinger (2003) says no:

> What would you say if someone came along and said, "Hey, we want to genetically engineer mentally retarded human infants! For reasons of scientific progress we need infants with certain cognitive and emotional deficits in order to study their postnatal psychological development." [...] You would certainly think this was not only an absurd and appalling but also a dangerous idea. It would hopefully not pass any ethics committee in the democratic world. However, what today's ethics committees *don't* see is how the first machines satisfying a minimally sufficient set of constraints for conscious experience could be just *like* such mentally retarded infants.

Obviously, the weight of this argument hinges to a large extent on the wide-open question of whether or not uploads are conscious, but here I do think Sandberg's (2014) Principle of Assuming the Most, discussed in Sect. 3 and serving as a kind of precautionary principle, should guide our decisions. And even if (contrary to our expectations) reliable arguments that computer consciousness is impossible would be established, we need not end up in Fukuyama's (2002) position that we lack moral reason to treat uploads well. We could for instance reason analogously to how Kant does about animals: he does not grant them status as beings to whom we can have moral duties, but claims nevertheless about man that "if he is not to stifle his human feelings, he must practice kindness towards animals, for he who is cruel to animals becomes hard also in his dealings with men" (Kant 1784–5, quoted in Gruen 2017). This kind of argument already appears in the current debate over permissible human treatment of sexbots (Danaher 2017).

The concerns about malfunctioning and suffering uploads raised by Bostrom and Metzinger above can be mitigated to some extent—but hardly eliminated—by experimenting with whole-brain emulations of animals ranging from nematode worms and fruit flies to mice and primates, before moving on to humans. This seems likely to happen. Still, assuming (plausibly) that destructive uploading comes first, the first humans to upload will be taking an enormous risk. While it would probably not be hard to find heroic volunteers willing to risk their lives for a scientific and technological breakthrough, the project is likely to encounter legal obstacles. We can probably not expect the law to grant personhood to uploads before the technology has already been established, whence at such a pioneering stage even an operationally successful upload will count legally as assisted suicide. Sandberg (2014) points out that "suicide is increasingly accepted as a way of escaping pain, but suicide for science is not regarded as an acceptable reason", and cites the Nuremberg code as support for the last statement.

Sandberg then goes on to suggest that the following might be way around this legal obstacle. Consider cryonics: the low-temperature preservation of legally dead people who hope to be restored to life in a future with far more powerful medical procedures than today's. The fact that cryonics is simultaneously (a) legal, and (b) a plausible hope for eventual successful awakening, is based on the (conjectured) existence of a gap between legal death and so-called information-theoretic death, defined as a state where a brain is so ramshackle that the person (his personality, memories and so on) can no longer be reconstructed even in principle: the necessary information is simply gone (Merkle 1992). If part of this gap can be exploited for uploading purposes by doing the destructive brain scanning post-mortem, then the legal (as well as the ethical) situation may be less problematic than the case of doing it to a living person.

Further ethical obstacles to uploading in the early stages of the technology concern the situation of an upload at that time, with no legal status as a person but stuck in a "legal limbo" (Sandberg 2014). At a later time, when uploading is more widespread, along with copying of uploads, legal and ethical concerns multiply. If I survive as two uploads, will just one of them be bound by contracts I have signed pre-uploading, or both, or neither? Who inherits my marriage, and who inherits my belongings? As to the latter, one might suggest that the uploads share them equally, but then what about inactive backup copies? Sandberg mentions all of these concerns (and others, involving vulnerability and privacy), as well as a challenge to our democratic system: surely uploads ought to count as full citizens, but then what if on election day I create 1000 copies of myself—does that give me 1000 votes? This brings us to the subject of the next section: what can we expect a society with widespread uploading technology to be like?

6 A Society of Uploads

A question of huge import to the issue of whether or not we ought to move forward towards an operationally successful uploading technology is what impact it is likely to have on society: what can we expect a society with uploading technology to be like, and would that be preferable to how society is likely to develop without uploading? Here, factual and normative issues are closely entangled, but in order to try to separate out the normative aspect, one may ask what kind of future society we want. A common answer is a society in which we have good lives, where "good" may for instance mean "happy" or "meaningful" (or some combination thereof)—terms that call for further definition. This quickly gets thorny, but there is a more concrete and well-defined wish that is at the focus in the growing research area of existential risk studies, namely the avoidance of extinction of humanity (Bostrom 2013; Häggström 2016; Torres 2017). To be able to lead happy or meaningful lives, we first of all need to avoid going extinct.

Sandberg (2014) mentions several ways in which a mind uploading technology may help reducing the risk for human extinction (provided that we decide that uploads qualify as human beings). One is that uploads are better suited for space travel than are biological humans, and so they facilitate space colonization in order not to "carry all eggs in one planetary basket". Another one is that if biological humans and uploads live side by side, a biological pandemic might wipe out the biological humans and a computer virus might wipe out the uploads, but it would take both events to exterminate humanity altogether.

Not going extinct is, however, far from a good enough specification of what might go into a desirable future for humanity. It has been suggested that, in the presence of suffering, survival could in fact be *worse* than extinction, and this has recently led to increased attention to a possible tension between actions meant to reduce the risk for human extinction and those meant to reduce the risk for creating astronomical amounts of suffering; see Althaus and Gloor (2016).

Hanson, in his *Age of Em: Work, Love and Life when Robots Rule the Earth* (2016a), which is the (so far) unchallenged masterpiece in the niche of non-fiction accounts of futures societies with uploads, summarizes more than two decades of his thinking on the topic. He avoids the normative issue almost entirely, settling instead for unsentimental predictions based on ideas from economics, evolutionary biology, physics and other fields. The main assumption he postulates is concerned with is that uploading technology becomes feasible before the creation of superhuman AGI, and before we attain much understanding of how thoughts and the brain's other high-level processes supervene on the lower-level processes that are simulated. The second part of the assumption prevents us from boosting the intelligence of the uploads to superhuman levels, other than in terms of speed by transferring them to faster hardware. The speedup becomes possible by what is perhaps Hanson's second most central assumption, namely that hardware will continue to become faster and cheaper far beyond today's level. This assumption also opens up the possibility to increase population size to trillions or more.

Hanson describes a society of uploads in fascinating detail on topics ranging from city planning, gender imbalance, management practices, surveillance and military operations to loyalty, status, leisure, swearing and sex. Changes compared to the way we live today are predicted to be in many ways large and disruptive.

Central to Hanson's scenario is the transformation of labour markets. The ease of copying uploads in combination with lowered hardware costs will push down wages towards subsistence levels: why would an employer pay you $100,000 per year for your work when an upload can do it at a hardware cost of $1 per year? Today's society in large parts of the world has been able to escape the Malthusian trap (for the time being), because ever since the industrial revolution we have maintained an innovation rate that has allowed the economy to grow much faster than the population. Our current evolutionarily maladaptive reproductive behaviour contributes to this gap, which Hanson notes is historically an exceptional situation. To emphasize its unsustainability (a view akin to the even darker but disturbingly convincing analysis by Alexander 2014), he calls our present era *dreamtime*. While a society of uploads will innovate even faster than ours, the ease with which uploads admit copying is likely to cause a population increase that the innovation rate cannot match.

Among the many exotica in Hanson's future is that uploads will be able to run on different hardware and thus at different speeds. This opens up the possibility to rationalize a company by removing some level of middle management and instead letting a single manager run at a thousand times the speed of the lower-level workers, giving him time to manage thousands of them individually; the lower-level workers can switch hardware and speed up in their one-on-one meetings with the manager.

Adjustable speed has the further advantage of making it easier to meet project deadlines, and so on.

An even more alien (to the way we live our lives today) aspect of the world painted by Hanson is the idea that most work will be done by so-called *spurs*, copied from a template upload in order to work for a few hours or so and then be terminated. Many readers have imagined the predicament of being a spur unbearable and asked why this would not trigger a revolution. Hanson (2016c) explains, however, that the situation of the spur is not much different from that of a person at a party knowing that he has drunk so much that by next morning his memories of the party will be permanently lost to amnesia. This strikes me as correct, at least if spurs accept the Σurvival = σurvival view discussed in Sect. 3—which they seem likely to do, as they feel psychologically contiguous with the template up to the time of copying.

More generally, readers of Hanson (2016a) tend to react to his future scenarios by describing them as dystopian. Hanson consistently rejects this label, maintaining that while the life of the uploads in his world may appear awful *to us*, it will actually be pretty good *to them*. Unfortunately, in Häggström (2016), I gave a misleading impression of Hanson's reasons for this, by suggesting that uploads "can be made to enjoy themselves, e.g., by cheap artificial stimulation of their pleasure centra". Such wireheading does not appear in his scenarios, and in Hanson (2016b) he corrects the misunderstanding, and gives his true reasons:

> Most ems work and leisure in virtual worlds of spectacular quality, and [. . .] ems need never experience hunger, disease, or intense pain, nor ever see, hear, feel, or taste grime or anything ugly or disgusting. Yes they'd work most of the time but their jobs would be mentally challenging, they'd be selected for being very good at their jobs, and people can find deep fulfillment in such modes. We are very culturally plastic, and em culture would promote finding value and fulfillment in typical em lives.

In his thoughtful review of *Age of Em*, Alexander (2016) criticizes the absence of wireheading and the relative sparsity of (the digital upload equivalent of) related pharmacological brain manipulation. Such absence can of course be taken as part of Hanson's central assumption that brain emulations are essentially black boxes so that we do not have the understanding required to usefully manipulate the inner workings of the emulations, but Alexander holds this assumption to be unrealistic if taken too far. Already today we can use amphetamine-based stimulants to significantly boost our ability to focus, and their main drawbacks—addictiveness and health concerns—become trivial in a world where, in Alexander's words, "minds have no bodies, and where any mind that gets too screwed up can be reloaded from a backup copy", whence

> from employers' point of view there's no reason not to have all workers dosed with superior year 2100 versions of Adderall at all times. I worry that not only will workers not have any leisure time, but they'll be neurologically incapable of having their minds drift off while on the job.

This leads Alexander to envision a world in which the economy as a whole keeps rolling with relentless momentum, but whose inhabitants turn into mindless puppets. Hanson (2016d) replies that the employers' incentive to use stimulants to boost their workforce may be tempered by the insight that "wandering minds may take away from the current immediate task, but they help one to search for hidden problems and opportunities", but the extent to which this fends off Alexander's dystopic logical conclusion seems to me uncertain.

Note finally that while this section has focused on a future with uploading technology, we also need to know something about the likely long-term outcome of a future *without* such technology in order to compare and to determine whether the technology is desirable. The latter kind of future falls outside the scope of this chapter, and I will just note that a key issue may be to figure out whether in such a future we have a way of avoiding the Moloch described by Alexander (2014).

7 Conclusion

I argued in Sect. 2 that mind uploading is likely doable eventually if we put sufficiently many decades of work into it. But should we do it? Sections 3–6 have featured a number of concerns that might cast doubt on that.

As to the concerns in Sects. 3 and 4 that uploading might be a philosophical dead-end, either by uploads not being conscious or by personal identity not surviving the uploading procedure, there seems to be a reasonable case for cautious optimism. Furthermore, regardless of what the true answer to these difficult questions is, one plausible scenario is that if uploading ever becomes a widespread technology, the philosophical concerns will quickly be forgotten. Uploads will consistently testify to us (and tell themselves) that they are conscious and that they are the same person as pre-uploading, and when these testimonials have become so commonplace as to seem superfluous, the philosophical concerns about uploading will perhaps eventually appear as silly as asking "Will I cease to be conscious if I have a cup of coffee?" or "Does taking a nap cause a person to be replaced by a different person who just happens to have identical personality and memories?".

Yet, even if these philosophical questions are settled in favour of survival under uploading (or if they are deemed ill-posed and therefore discarded), there still remains the difficult bioethics-like concerns discussed in Sect. 5, as well as the question discussed in Sect. 6 on whether uploading technology is likely to lead to a society worth wanting. Whatever knowledge we have on these issues at present is highly tentative, and it is important that we put serious efforts into resolving them, rather than merely pressing ahead blindly with technology development. The future of humanity (or posthumanity) may be at stake.

Acknowledgments I am grateful to Dorna Behdadi and Björn Bengtsson for helpful discussions.

References

Alexander, S.: Meditations on Moloch. Slate Star Codex, Jul 30 (2014)
Alexander, S.: Book review: age of Em. Slate Star Codex, May 28 (2016)
Althaus, D., Gloor, L.: Reducing Risks of Astronomical Suffering: A Neglected Priority. Foundational Research Institute, Berlin (2016)
Bach, A.: The concept of indistinguishable particles in classical and quantum physics. Found. Phys. **18**, 639–649 (1988)
Bach, A.: Indistinguishable Classical Particles. Springer, New York (1997)
Baker, A.: Simplicity. In: Zalta, E. (ed.) The Stanford Encyclopedia of Philosophy (2010)
Bergström, L.: Recension av Here Be Dragons. Filosofisk Tidskrift. **4**, 39–48 (2016)
Blackford, R., Broderick, D.: Intelligence Unbound: The Future of Uploads and Machine Minds. Wiley Blackwell, Chichester (2014)
Blackmore, S.: She won't be me. J. Conscious. Stud. **19**, 16–19 (2012)
Bostrom, N.: Existential risk prevention as global priority. Global Pol. **4**, 15–31 (2013)
Bostrom, N.: Superintelligence: Paths, Dangers, Strategies. Oxford University Press, Oxford (2014)
Bostrom, N., Savulescu, J.: Human Enhancement. Oxford University Press, Oxford (2009)
Chalmers, D.: The Conscious Mind. Oxford University Press, Oxford (1996)
Chalmers, D.: Uploading: a philosophical analysis. In: Blackford, Broderick, pp. 102–118 (2014)
Danaher, J.: Robotic rape and robotic child sexual abuse: should they be criminalised? Crim. Law Philos. **11**, 71–95 (2017)
Dennett, D.: The myth of the computer: an exchange, New York Review of Books, Jun 24 (1982)
Dennett, D.: From Bacteria to Bach and Back: The Evolution of Minds. W.W. Norton & Co, New York (2017)
Freitas, R.: Nanomedicine, Vol. I: Basic Capabilities. Landes Bioscience, Georgetown, TX (1999)
Fukuyama, F.: Our Posthuman Future: Consequences of the Biotechnology Revolution. Farrar, Straus and Giroux, New York (2002)
Grace, K., Salvatier, J., Dafoe, A., Zhang, B., Evans, O.: When will AI exceed human performance? Evidence from AI Experts. arXiv, 1705.08807 (2017)
Gruen, L.: The moral status of animals, The Stanford Encyclopedia of Philosophy, Fall 2017 Edition (ed. E. Zalta) (2017)
Häggström, O.: Here Be Dragons: Science, Technology and the Future of Humanity. Oxford University Press, Oxford (2016)
Hansell, G., Grassie, W.: H+/-: Transhumanism and Its Critics. Metanexus Institute, Philadelphia (2011)
Hanson, R.: Age of Em: Work, Love and Life When Robots Rule the Earth. Oxford University Press, Oxford (2016a)
Hanson, R.: Here be dragons. Overcoming Bias, Jan 16 (2016b)
Hanson, R.: Is forgotten party death? Overcoming Bias, Apr 23 (2016c)
Hanson, R.: Alexander on age of Em, May 30 (2016d)
Hofstadter, D., Dennett, D.: The Mind's I. Basic Books, New York (1981)
Kant, I.: Moral philosophy: Collin's lecture notes. In: Heath, P., Schneewind, J.B. (ed. and trans.) Lectures on Ethics (Cambridge Edition of the Works of Immanuel Kant). Cambridge University Press, Cambridge 1997, pp. 37–222. Original is Anthropologie in pragmatischer Hinsicht, published in the standard Akademie der Wissenschaften edition, vol. 27 (1784–5)
Kurzweil, R.: The Singularity Is Near: When Humans Transcend Biology. Viking, New York (2005)
Lem, S.: Dialogi, Wydawnictwo Literackie, Krakow. English translation by Frank Prengel at LEM. PL. http://english.lem.pl/index.php/works/essays/dialogs/106-a-look-inside-dialogs (1957)
McGinn, C.: Consciousness and Its Objects. Clarendon, Oxford (2004)
Merkle, R.: The technical feasibility of cryonics. Med. Hypotheses. **39**, 6–16 (1992)
Metzinger, T.: Being No One. MIT Press, Cambridge, MA (2003)

Moravec, H.: Mind Children: The Future of Robot and Human Intelligence. Harvard University Press, Cambridge, MA (1988)

Parfit, D.: Reasons and Persons. Oxford University Press, Oxford (1984)

Pigliucci, M.: Uploading: a philosophical counter-analysis. In: Blackford and Broderick, pp. 119–130 (2014)

Sandberg, A.: Ethics of brain emulations. J. Exp. Theor. Artif. Intell. **26**, 439–457 (2014)

Sandberg, A., Bostrom, N.: Whole brain emulation: a roadmap. Future of Humanity Institute technical report \#2008-3 (2008)

Schwitzgebel, E.: The crazyist metaphysics of mind. Aust. J. Philos. **92**, 665–682 (2014)

Searle, J.: Minds, brains and programs. Behav. Brain Sci. **3**, 417–457 (1980)

Searle, J.: The myth of the computer. New York Review of Books, Apr 29 (1982a)

Searle, J.: The myth of the computer: an exchange (reply to Dennett). New York Review of Books, Jun 24 (1982b)

Searle, J.: Mind: A Brief Introduction. Oxford University Press, Oxford (2004)

Sharvy, R.: It ain't the meat it's the motion. Inquiry. **26**, 125–134 (1985)

Torres, P.: Morality, Foresight, and Human Flourishing: An Introduction to Existential Risks. Pitchstone, Durham, NC (2017)

Yudkowsky, E.: Timeless identity. Less Wrong, Jun 3 (2008)

Yudkowsky, E.: There's No Fire Alarm for Artificial General Intelligence. Machine Intelligence Research Institute, Berkeley, CA (2017)

Chapter 2
Transhumanism as a Derailed Anthropology

Klaus Kornwachs

Abstract According to some proponents, artificial intelligence seems to be a presupposition for machine autonomy, wheras autonomy and conscious machines are the presupposition for singularity (Cf. Logan, Information 8: 161, 2017); further on, singularity is a presupposition for transhumanism. The chapter analyses the different forms of transhumanism and its underlying philosophical anthropology, which is reductionist as well as naturalistic. Nevertheless, it can be shown that transhumanism has some (pseudo-)religious borderlines. Besides this, massive interests behind the arguments of the proponents can be figured out. Due to these hidden business models, it would be a good idea to discuss objection and rules to hedge an uncontrolled shape of these technologies.

1 Introduction: The Story so Far

It is a very characteristic trait in all cultures and epochs that mythologies, religious belief systems with relevant theologies or ideologies like historicism, Marxism as well as neo-liberalism are used to refer to frames going beyond the existence of individual man and the history of mankind. Moreover, such ideas try to go beyond time and space, and—speaking in modern terms—beyond the imagination what the fate of mankind could be. It is a common trait of questions about transcendence: Where do we come, where will we go? Looking at answers, some common elements are revelation, salvation and redemption, whereas the distinguishable preconditions of those belief systems are very special: They differ mainly in their concrete anthropologies, i.e. in their views of the role of human entities in natural or salvation history, in society, in the world, in environment, in cosmos and so on. Many of those—frequently anthropocentric—approaches show quite different solutions. For

K. Kornwachs (✉)
Humboldt Center for Humanities, University of Ulm, Ulm, Germany
e-mail: klaus.kornwachs@uni-ulm.de; klaus@kornwachs.de

W. Hofkirchner, H.-J. Kreowski (eds.), *Transhumanism: The Proper Guide to a Posthuman Condition or a Dangerous Idea?*, Cognitive Technologies,
https://doi.org/10.1007/978-3-030-56546-6_2

instance, the difference between animal and man is outlined in many ways.[1] Other philosophical battlefields are represented by the mind–body problem with differently discussed topics in terms of monism and dualisms, and by the issue of how the gap between religious views and evolution theory could be bridged.

The so-called three big insults of human self-understanding, known by the names Copernicus, Darwin and Freud,[2] have been now completed by the fourth offence: Machines are or will be superior to a man with respect to calculation ability, capacity and velocity. The angry discussion about "*what computer can't do*"[3] can it be really understood as a rearguard action in favour of anthropologic positions of the nineteenth century?

Within the last few years, it has been claimed that one day we may develop a technology which could show a certain degree of autonomy such that these technologies do not need us anymore. This is the thesis of singularity.[4] But the key point is highly questionable: For what and whom such an autonomous technology is needed and for what purposes machines do not need us anymore? Is it an end in itself? This would presuppose that the machinery may be able to set its own goals with no external influence. In other words, the machine will be able to develop its own and genuine intentions[5] such that utility or profit functions can be defined by the technology as an end in itself.

The idea is not very new: Robots as autonomous, humanoid entities, i.e. with defined borders between themselves and other entities or environment, as well as with the ability of self-replicating and self-improving, will start someday a new evolution. The steps

A. Coexistence with the human race
B. Melting of the human race and robot race (cyborgs)
C. Superiority over mankind
D. Conflict with the human race, slavery
E. Extinction of the human race

---

[1]This is marked in the so-called anthropological constants like self-consciousness, consciousness of own mortality, self-critique, language, laugh, love and hate, ironic and metaphoric communication, free will, historic memory (Landmann 1982) as well as curiosity, imagination, intuition, emotions, passion, desires, pleasure, aesthetics, joy, purpose, objectives, goals, télos, values, morality, experience, wisdom, judgement and even humour (cf. Braga and Logan 2018).

[2]Copernicus destroyed the Ptolemaic worldview, and Kepler showed that the same physics holds in heaven and on earth; Darwin showed that man is a product of the evolution process, and Freud demonstrated the power of the un- and non-consciousness.

[3]Cf. Winograd and Flores (1986) and Dreyfus (1992). With respect to arts, cf. Kelly S. (2019) who denotes painting, music composing, poetry, etc., done by computer programs, as mimicry—all this has nothing to do with creativity.

[4]Cf. Kurzweil (1999, 2006) and Joy (2000).

[5]Intention has two meanings: Within the context of phenomenology, the intention is defined e.g. by Husserl (1970) as a basic property of consciousness, to be directed to an object (the believed, the wanted, etc.). In psychology intention is quite simply the motivation to act, intentional stance, according to Dennett (1987).

have been discussed already in science fiction literature as well as by scientists and in philosophy.[6] For all paths of these thinkable developments, we have not yet posted the question: What is the benefit for man and mankind?

Convergence of technological systems and their organizational closures,[7] well equipped with AI, could generate a distributed, delocalized worldwide system of technology. Such a system could be empowered to make more and more worldwide decisions in favour of the interest of the developers, owners and operators of the system. These decisions are only relevant with respect to certain value chains; e.g. security systems will converge up to social control systems.[8] The interconnectivity may allow a kind of evolution of such a hyper-system whose decisions cannot be controlled anymore by the man with respect to time, complexity and irreversibility. Humans become superfluous step by step with respect to the value chains. Nevertheless, how the term "value" can be defined in the context of a process, called value chain which is detached from human life and the living conditions of man?

This gives rise to the hypothesis that such future visions are inspired by a certain image of man the protagonists of singularity and transhumanism have. First of all, it is tried to suggest to the layman that such a development is inevitable as a kind of historical necessity.[9] Second, the protagonists are themselves deeply engaged in the development of AI and robotics.[10] Nevertheless, their promises of a brave new world with the help of AI and robotics can be valid only for the time period in which mankind can enjoy the blessings of these technologies up to the point when machines start to define their own purposes. This obvious irrationality of promises must have a psychological source, and the conjecture will be here that this source is based not on a religious image of man but of a pseudo-theological transfiguration of the machine as such.

2 Manifestations of Transhumanism

2.1 Three Types of Transhumanism

Within the history of ideas about transhumanism, three types can be distinguished:

[6]See Bostrom (2005).

[7]We can observe convergence processes between separate development lines of technologies. Digitalization can be understood as a convergence between analog and digital technologies, i.e. between telecommunication and computer technology. Further convergence has happened or will happen between biology, neuro-science, artificial intelligence, nano-science, cognitive science, etc.; discussed in Roco and Bainbridge (2002) and Roco et al. (2013).

[8]See the Chinese social score systems, cf. Botsman (2017).

[9]Bostrom (2005), Braga and Logan (2018), Moravec (1990), and Fuller and Lipinska (2014).

[10]Kurzweil is Research Director at Google, Minsky, and Moravec had chair for AI, etc. at MIT.

1. Extropians: Due to a technology of enhancement, genetic engineering and improvement in all living conditions, the evolution of man can be accelerated towards a superior man with respect to information processing and storage capacities as well as "*intelligence, ... available energy, longevity, vitality, diversity, complexity, and capacity for growth*".[11] These visions encompass step B mentioned above.
2. Singularity: The human species will be transcended by human-made super-intelligent machines (or robots) which can act autonomously, which can reproduce and improve themselves and which will rule earth and space due to the fact that they will be superior with respect to all human properties (step C). Here, singularity is a presupposition for such kind of transhumanism,[12] which may include the steps D and E.
3. Point omega or noosphere: Due to the rise of human cognition, the surface of the earth has been changed, comparable to the change of the chemical conditions on earth after the rise of biological systems. Human cognition leads to a growing density of communication and connectivity between individuals. Due to this effect, it seems to be possible that the individuals get the same role in biosphere as the cells within an organism. Thus, a completely new organism will emerge, possibly with its own cognitive status, i.e. with self-consciousness and intentions. This would be the evolutionary end of human individuals. There are several variants of this hypothesis like the noosphere of Teilhard de Chardin with some theological grounding.[13] Thus, the Gaia hypothesis assumes an already existing regulating ecological meta-system on earth.[14] But all these variants remain at least on step C.[15]

For all these types and variations thereof, there are ancestors in the history of ideas. Not necessary to mention the myth of Prometheus, who was punished by the gods not only because he has stolen the fire from the Olympus (and other misdoings), but also due to his experiment to make his creations better than man has been created by the Gods. To expand human boundaries means to acquire eternal youth, to get access to a panacea, to become able to force the nature according to the will of man—all these topics can be found in myths and tales in all former ancient cultures up to the medieval alchemists and their attempt to create the homunculus.[16] Up to this time, the mirroring of man in his own creation and the enhancement of man, opposite to nature, is a predominant motivation. This can also be found in the above types (1) and (2).

---

[11]More (1993, p. 1).

[12]Transhumanist declaration in https://humanityplus.org/philosophy/transhumanist-declaration/

[13]Chardin (1964/2002).

[14]Lovelock (1998) and Margulis (1999).

[15]See, e.g., Tipler (1994) and More (1993).

[16]Bostrom (2005).

2.2 *Future Paradise as a Kind of Transhumanism?*

Type (3) has some religious and some secular forerunners. In the book *Apocalypse*, after doomsday, and after a millennial reign a new man and a new world are propagated.[17] Comparable chiliasm can be found in Hinduism, Judaism and Islam.[18] A non-apocalyptic transformation of man into new mankind has been given by Teilhard de Chardin with his theory of point omega:[19] Due to the increasing intensity of communication between all living individuals (which is now thinkable due to actual communication technologies), the single consciousness is melted to a kind of thinking, conscious organism called noosphere, in which individuals play the same role as cells in a neuronal net. This is the dissolution of individual person into an organism and its collective consciousness which is much wiser than individuals ever could be. This dissolution of the individual personality as the source of evils and sins is the process of the promised salvation. Chardin has interpreted such a development as a point omega, which is identical to the rule of Christ.[20]

Without a religious background, the Russian biochemist and geologist Vernadskij argued the chemical and biological transformation of the earth's surface by living and thinking entities.[21] He proposed a development scenario for another kind of noosphere, based on natural science.[22] After the geosphere, consisting of an-organic, nonliving matter, and the biosphere with an animated matter, there could be a kind of noosphere: After the rise of life, the chemical properties of geosphere have been transformed drastically. The rise of human cognition will drastically change the biosphere; therefore, Vernadskij assumed that there will be a further evolution on a cognitive level. This idea has been adapted later e.g. by Teilhard de Chardin. This should not be confused with the concept of Global Brain[23]: This hypothesis assumes that the already existing information and communication technology network tends to connect more and more humans and their devices. Due to this network, a certain swarm or collective intelligence will emerge, taking over more and more function of communication and organizational governance. Therefore, it could be interpreted as a Global Brain which will become the Brain of the Earth.[24]

---

[17]Apocalypse 20:1–4; 21.

[18]Analogous ideas can be found in Jewish religion, cf. Daniel 2:44. For Islam, cf. Quran 14:45–53; 75:4–17; cf. also Madelung (1986). For Hinduism cf. Glasenapp (1996).

[19]Chardin (2002).

[20]Cf. Ratzinger Pope Benedict XVI (2006), Kindl Location 260–270. See also Pfleger (1964, p. 129 ff).

[21]Vernadskij (1997) and Vernadsky (1997).

[22]See Levit (2000), Vernadskij (1997), and Vernadsky (1997).

[23]E.g. Heylighen and Lenartowicz (2017).

[24]See Heylighen and Bollen (1996).

2.3 Marxism

A more secular idea of a future paradise has been coined by Karl Marx. Man is creating himself by work, and he is changing the nature and the nature of himself.[25] Due to history as a series of class fights the economic conditions, having changed the consciousness of man, and the societal structure will turn into a classless society without exploitation, suppression and strenuous work.[26] Marx's categorical error was to confound the formal concept of nature (nature of man) and the material concept (nature as all naturally generated objects around us). Today, man is considered as a part of nature, not an opposite entity. He cannot change the nature in its physical structure, but he can change his environment by collecting, hunting, burning and grubbing forests, ordering his fields, building machines, organizing societies, making arts and producing civilizations. According to Leibniz's dictum: A stone is not yet a plant, a plant is not yet an animal, the animal is not yet a man, Ernst Bloch asked in a talk: And the man—is he not yet a man?[27] But which man?

In nearly all religious books, the redemption of man is an important part of the message. There are two variants that also occur in mixed ways:

1. There are the redemption of burdens and suppression by a future rule, sometimes expressed as a millennial reign. This is called chiliasm, and the future reign can be interpreted either politically or as a metaphor for a hereafter heavenly kingdom.[28]
2. There is a personal life after physical death, a real disembodied survival of the individual with all its characteristics, subjectivities and memories. All sins are forgiven and this life will be happy and eternal.[29] We will find this second variant in the version of the extropians (see above) and the possibility to upload the brain contents to a system in silico.

The Marxian self-creation of man due to labour belongs to the first variant (1) above. The experiment of communism failed at least in 1989 since the nature of man could not be changed only by suitable calibration of economic conditions as Marx believed. Nevertheless, the belief that man can enhance himself, i.e. to extend his abilities by technology, has been proved to be true.[30] This positive experience has amplified the technologic efforts, and man has indeed changed his environment

[25]"*(indem der Mensch) auf die Natur außer ihm wirkt und sie verändert, verändert er zugleich seine eigene Natur*". Das Kapital I, MEW 23 Kap. 5, S. 192, engl.: *By thus acting on the external world and changing it, he at the same time changes his own nature.* Cf. Marx (1990): The Capital I, Part III, chap. 6, Sec 1.

[26]Even Thomas Moore has described 1516 in his novel *Utopia* a society free from labour.

[27]Ernst Bloch: Talk at Leibniz-Kolleg, University of Tübingen, December 1966, referring to Leibniz (1982).

[28]Variants can be found in Jewish Torah, in Christian New Testament and in Quran, as well as in the writings of Hinduism and Jainism.

[29]This holds at least for Christian and Islamic religions.

[30]Cf. early Grunwald (2007).

up to a dangerous limit: Technology has proceeded so far that it is possible to undermine the conditions for the possibility of technology at all by extinction of mankind (e.g. climate change, nuclear weapons, biologic hazard and environmental pollution).

There are some correspondences between Marx and Christian religion: The end of mankind is also the end of the world known as a familiar environment, and this apocalypse is the very beginning of a new world and of a new form of man's existence. As mentioned above, the palaeontologist and theologian Teilhard de Chardin has interpreted this development as a trend to the point omega, where all individuals share the same thoughts due to an intensive interconnection. Mankind will become a new organism with a new conscious.[31] De Chardin could not have any knowledge about today's Internet and networking, but he conceptualized a new nature of man and a new world in the framework of a radical Christology.[32]

2.4 Gaia Hypothesis

A secularized expression of such hopes can be found in the so-called Gaia hypothesis.[33] Gaia is the name of an ancient goddess of earth in Greek mythology. This idea assumes that the biosphere of our planet forms a living organism, which is a symbiosis of all organisms, enabling live and evolutionary development. Thus, an organic entity, let it be conscious or not, is hypothesized, being able to regulate material processes on earth by controlling and feedback mechanisms. Man is part of this organism like a cell within a biological body. Gaia performs a genuine and necessary development in order to improve its own conditions of existence. Thus, the fate of mankind depends upon its behaviour, i.e. upon the degree of being adapted to the goals of Gaia.[34] Some later adoptions of the Gaia hypothesis have abandoned the natural science base and assume an animated conscious overall system, which is able to punish our eco-sins and to reward measures to save the world.[35] In a more weak version, Gaia can rule some material processes on earth with the purpose of stabilizing the ecosphere. Nevertheless, all versions use metaphors with unclear meanings and rather conjuring functions.

The section above shows that transhumanism is not an invention of AI proponents, but an adaption of more or less religious or mythical images in the light of new technological possibilities. Thus, as a first result, transhumanism tries to make forecasts for future developments in terms of promises. Quite the opposite, the

[31] Chardin (2002).

[32] Pfleger (1964).

[33] Lovelock (1998) and Margulis (1999).

[34] Ibid.

[35] A critique is given by Trepl (2005). There can be a herd of healthy sheep, but there is no healthy herd of sheep (according to Ernst Mayr, cited in Trepl (2013)). See also Dawkins (1982).

teachings of evolution theory refrain from making any forecasts, but they can explain the past development in terms of necessity and chance. There is neither teleology nor theology in it.

3 Anthropology and the AI Image of Man

3.1 Man, Animal, Machine

Usually in philosophical anthropology, the specification of possible answers to the Kantian question "*Was ist der Mensch?*" [36] is given by ruling out the distinction between man and animal. This distinction can be characterized by so-called anthropological constants as mapped in Table 2.1, according to the literature in anthropology.[37]

In occidental metaphysics of rationality and enlightenment, man is used to be defined as the rational animal (*animal rationale*). In this definition, man remains an animal with some additional features. This may be regarded as the beginning of a reductionist program of scientific explanation: Man has to be conceptualized primarily from a biological point of view, whereas the other properties mark the decisive difference between man and animal. Anyway, they seem to remain as something extra. The philosophy of German idealism has pointed out that transcendent or spiritual issues are not additional, i.e. contingent but essential for any understanding of man. Man is in his existence (*Dasein*) within the world, not only within an environment[38] like dogs and plants. Man communicates with other men and he is able to establish a realm of communication by reflecting on the constraints of this realm. Thus, he can willingly support or destroy the conditions for the possibility of communication, and he can understand and last but not least he is able to shape a society.

Provoked by the discussion about "*What computers still can't do*",[39] similar "factors" have been eroded, also listed in Table 2.1. This discussion is ongoing, and it has some features of a war of faith. The third relation within the triangle man—machine—animal has been rarely discussed. This is the distinction between machine and animal. Although the biologization of technology is actually an issue, the technicalization of biology is quite an old process, from the domestication of cows, horses and dogs up to the use of virus and bacteria to change work material

[36] Kant's first three fundamental questions "*Was sollen wir tun?*" .... "*Was können wir wissen?*" .... "*Worauf sollen wir hoffen?*" can be found *in Critique of Pure Reason*, A 805 rsp. B 833, cf. Kant (1995, IV, p. 677). The last question "*Was ist der Mensch?*" can be found in Kant's Logik (A 25) in Kant (1995, VI, p. 448).

[37] Inspired by e.g. Gehlen (1957), Landmann (1982), and Braga and Logan (2018).

[38] Heidegger (1986, §26) uses the term: *in-der-Welt-sein* (eng.: *being-in-the-world*).

[39] Dreyfus (1992).

Table 2.1 Distinguishing factors between man, animal and machine

Tertium comparationis	Man	Biological entity	Machine
Transcendent abilities	*Reflexivity*: self-consciousness, free will, the consciousness of mortality, self-critique, historical curiosity, self-reflection about the meaning of existence	Not known	Not known Not possible on von Neumann machines due to self-referential expressions
Cultural abilities	*Thinking*: justifying, the division of work, to form communities, making science, organization and technology *Imagination*: intuition, arts, pleasure, aesthetics, joy	Division of work	Any decision, which machine should be connected with other ones is implicitly done by the network installer
Social abilities	*To form systems (institutions)*: moral systems and economic systems *Expression* of emotions (see below)	Symbiotic communities Swarm intelligence	Not known Connected machines can show simulated swarm intelligence
Emotional abilities	Having emotion: anxiety, fear, yearning, sense of community, affections, empathy, pity, charity, greed, competitiveness, revenge and forgiveness, love, hate, jealousy, passions, desires *Suppression* of pain and discomfort	Anxiety, fear, sense of community, affections, empathy, greed, competitiveness	Simulation possible within defined contexts (critically: Lunceford (2018))
Cognitive abilities	*Self-Awareness*: rational and logic thinking, having doubts *Judgement*: experience, decision, wisdom	Partially consciousness (higher vertebrates), memory	Self-sensory state control, deductive and abductive procedures (finite class logic and restricted, i.e. decidable and complete modal logic)
	Communication: language, ability to perform intentionally speech acts, ironic and metaphoric communication, laugh, humour, the difference between to mention and to use	Ability to perform intentionally speech acts like to warn, to menace, to attract, to lure	No intentionally acts, but natural language processing

(continued)

Table 2.1 (continued)

Tertium comparationis	Man	Biological entity	Machine
	Genuine goals and meta-goals, intentions, motivations	Instinctively generated goals, intentions	Programmed, i.e. externally given goals and meta-goals
	Orientation in space and time	Orientation in space and time	Orientation in space and time
	Learning	Learning	Learning
Moral abilities	Morality: objectives, goals, télos, values, purpose	Not known	Not yet known
Information processing	Slowly but massive parallel information processing	Slowly but massive parallel information processing	High-speed sequential processing (incl. Emulated neuronal nets)
Abilities	Pattern recognition	Pattern recognition	Pattern recognition
Instinctive abilities	Partially instinctive (i.e. not rationally reflected) reaction on unexpected situations	Overall instinctive reactions	Programmed non-reflective behaviour
Skills of action /Sensory perception Capabilities	Sensory perception (optic, haptic, olfactory, smelling, acoustic)	Sensory perception + electromagnetic sensitivity in many reaches better than man	Across the electromagnetic and acoustic spectra partially better than man
	Actions of hand	Paws and cross tools	Artificial hands
	Walking Dancing	Moving in all forms	Restricted, a defined reach of motion
Self-reproduction	Basic	Basic	Not known

properties or to dissolute organic or even inorganic debris. Taking into account the distinction between biological entities and machines also may tell us about the nature of the machine and the nature of man.

Thus, we can recognize from—anyway imperfect and incomplete[40]—Table 2.1 that the difference between machine and biological system is given in the social, emotional and cognitive abilities as well as in the way of information processing. This holds at least for the higher species. The purpose of this table is to show that we are even far away to build genuine biological systems. This does not exclude the possibility to modify already existing biological systems according to our wishes by

[40]Voss (2017, § 12) discusses eight properties a machine should have; it may be comparable to human cognitive ability, like to learn something new from a single example, be able to reason contextually, logically and abstractly, to understand the context and purpose of actions, including those of other actors, to use existing knowledge and skills to accelerate learning, to form abstractions and ontologies, to dynamically manage multiple, potentially conflicting goals and priorities, to recognize human emotions and its own cognitive states, to handle all that with limited knowledge, computational power and time.

gene technology, breeding, synthesis, cultivation and so on. But it has not yet been possible to build a living machine from nonliving components.

All these circumstances give rise to the hypothesis that any philosophical anthropology has to take into account these differences as pivotal. In other words, the reductionist view (see below) that the differences between man, machine and living systems can be removed, i.e. explained by a unified concept of the machine, i.e. algorithmically expressed processes, has the burden of proof to show this unified concept. This has not yet been done to the knowledge of the scientific community.

3.2 Man Embedded in Systems

In the discussed technologies like AI, robotics and autonomous systems, etc., the fundamental question of philosophical anthropology arises: Is man still autonomous if he is embedded within such systems which he himself has created for his own purposes? What is up with his autonomy if he starts to allow the technical systems to modify themselves, to adapt or even to create other machines like themselves? Moreover, who decides on the decision criteria of fully automated systems of this second degree? How many levels of the criteria still remain with the designing and building man? Which criteria are already machine learnable or delegable to machines?

Already in this context, there is an intensive discussion on which image of man the design of autonomous weapons, the analysis of personal data and the interpretation of their evaluation in personnel management are based on.[41] Which type of person is preferred to work in Industry 4.0? Is the human being seen by the system providers only as the high performer, as an actuator in a control system, as a self-optimizer, as a cost factor, as a *meat machine*?[42] Going a step further, one may pose the question within the context of gen-technology and the possibility of influencing the germ line: Who decides according to which criteria in what direction the biological and genetic future of mankind should be driven?[43]

There is—feared by many—a connection between the—at least propagated—technology development of computerization, networking and replacement of human work processes and services on the one hand and the emergence of strong and harsh critiques of democracy and economics on the other hand, together with the moral dissolution of the previous political rules of conduct by populism. These ideas present themselves in a wide range of political movements and also in the fantasies of the world improvers à la Silicon Valley. This connection results from a changed

[41]Kornwachs (2018).

[42]Marvin Minsky: "*The brain happens to be a meat machine.*" Cit. from Weizenbaum (1976/1993, pp. 72–73).

[43]Sloterdijck (1999) has provoked such a debate in his speech: "*Regeln für den Menschenpark*"/ *Regulation for the Human Park*. See also Mewes (2002).

view of the man whose contours are still unclear, which—nevertheless—seems to be increasingly technically determined.[44]

There are two fallacies that may have accelerated this development, or even have made it possible: the socially determined perception and thus also the theoretically conditioned and interest-guided comprehensions of reality were the starting points for a fundamental error of postmodernism philosophy: Even scientific facts (brute facts) have been considered to be exclusively socially constructed. Thus, even physical theories degenerated into narratives.[45] If everything is only a narrative, concepts of truth and ontological models may be dissolved into arbitrariness and interests. This could just be accepted since these critiques could apply to economic and sociological contexts that relate to institutional facts, i.e. constituted by human agreements. Physics is not dealing with institutional facts, but with natural (brute) facts.[46] However, the current effect is that there is no unanimous communication about facts in some parliaments.

The second fallacy is to consider technology as neutral in value, and—as a consequence—everything that is connected with artificial intelligence and robotics is seen as purely technical, logical and rational. But this position disguises the interests and business models which the builders of such products actually may have or will have in the future. Anyone who invents, designs, produces, processes and disposes of technology has interests. That is trivial but mostly forgotten.

Thus, there is very often a lack of criteria for determining what a fact is and what may be true. If we do not recognize, we cannot control. If these systems, which are seemingly neutral, only user-dependent, are a carrier of interests, the path is ready for the confusion between what we have manufactured by human efforts and what is natural, i.e. produced by natural processes, and not influenced by human interaction.

These questions touch upon the fundamental question of philosophical anthropology: Is man still autonomous or even free (according to Kant) if he is embedded in such seemingly autonomous artificial systems of life-world—having no criteria to distinguish between fakes and facts or the natural and the artificial?

4 Autonomy as a Presupposition for Singularity

The usual AI paradigm is seen as a conjecture: It is possible to find algorithmic processes, which can be implemented on Von Neumann machines (or future modifications thereof like quantum computer, dedicated neuronal nets, synthesis of biological tissues and digital circuits, etc.), in order to generate cognitive processes

[44] An early critique has been given by Boden (1977).

[45] Lyotard (1979, 1988).

[46] The difference has been shown by Popper; cit. acc. to Popper (1992, p. 70–71): You cannot take out more coins than are in the burse. This is a natural or brute fact. But you can cover your account if you have negotiated it with your bank. This is an institutional fact.

up to higher abilities in Table 2.1. Besides this strong AI paradigm, a weaker paradigm states the possibility of—more or less perfect—simulations of such processes and abilities.

Autonomy can be defined in analogy to the discussion of autonomous weapons in three stages[47]:

(a) In-the-loop (IL): non-manned devices (missiles, robots, etc.), but programmed, operated and controlled by man.[48]
(b) On-the-loop (OL): programmed and controlled by man, operating automatically, partially in modus of decision substitution.
(c) Out-of-the-loop (ExL): Self-programming systems, operating autonomously and self-controlled with implicitly programmed meta-goals.

For autonomous systems in stages, weak AI is a necessary but not sufficient presupposition, for systems in stage ExL the strong AI paradigm could be a presupposition. Whether this requirement is a necessary condition or not cannot be answered yet. For singularity, i.e. the development of systems surmounting the human abilities in Table 2.1, an autonomy stage of systems, comparable with ExL, would be necessary. As a consequence of this classification, man would be forced to refrain from the control of such systems.

To proceed, we can ask whether singularity can be thought a presupposition for transhumanism. In a weak version, the extropians (1), not even strong AI is needed. For the version of noosphere (3), strong AI may play a role, but not necessarily. Transhumanism inversion (2), due to singularity, needs strong AI, since the epistemic claim of strong AI, together with belief in continuous technological progress, will lead to singularity and as a further step to transhumanism.

We can suspect that the chain of argumentation covers some hidden interest. The first indication is the timely coincidence between the discussions about the alleged non-existence of free will and the mind's I as an illusion[49]and a naturalistic view of man as a "*meat machine*" on the one side and on the discussions about strong AI, singularity and transhumanism on the other side. If this coincidence is not accidental, we could guess some interests behind the arguments of the proponents of these discussions.

---

[47]Misselhorn (2018, p. 157 ff.), ref. to the United State Department of Defense (2011).

[48]Not really autonomous, but in public discussions about weapons, unmanned missiles are also called autonomous.

[49]Libet (1999), Roth and Strüber (2017), and Singer (2004a, b).

5 Interests, Concrete and Future

5.1 Religious Motivations and Interests in Transcendence

In his *Oration on the Dignity of Man*, which has been called the "Manifesto of the Renaissance", and which can be considered as a key text of Renaissance humanism. Giovanni Pico della Mirandola characterizes his anthropology, "citing" the sayings of God when the creator put Adam into existence:

> (19) "The nature of all other creatures is defined and restricted within laws which We have laid down; (20) you, by contrast, impeded by no such restrictions, may, by your own free will, to whose custody We have assigned you, trace for yourself the lineaments of your own nature. (21) I have placed you at the very center of the world, so that from that vantage point you may with greater ease glance round about you on all that the world contains. (22) We have made you a creature neither of heaven nor of earth, neither mortal nor immortal, in order that you may, as the free and proud shaper of your own being, fashion yourself in the form you may prefer. (23) It will be in your power to descend to the lower, brutish forms of life; you will be able, through your own decision, to rise again to the superior orders whose life is divine."[50]

This section shows clearly the spectra of human striving under the condition of free will. It allows us to surmount the usual *conditio humana* towards a superman as well as to be reduced to a functional part of a herd of animals. This manifesto has defined for a long time the self-understanding of man up to the Marxism[51] and existentialism of the twentieth century: Man is what he defines by will and action.[52]

The way to bridge the gap between this kind of thinking and transhumanism at least type (1) up to the step (B–C) is easy. It may be of interest to surmount the human nature by will to power,[53] by the revolutionary change of economic conditions (Marx) and by the development of enhancing technologies to create a new world and a new man. It is therefore of interest to put forward the development willingly (as a cultural effort), or to develop better social and economic structures (like in post-Marxism) or to put forward the technologies proactively like in Silicon Valley.

5.2 Economic Interests

Beyond hidden religious or transcendence interests, there are obvious hard interests of the protagonists of singularity. Supposed that AI development is a presupposition of singularity as mentioned above, and presupposed, that singularity will free us

[50]Pico della Mirandola, Giovanni: Oratio de hominibus dignitate (1496, §5, 19–23).

[51]Man is creating himself (i.e. his own nature) by work. Cf. Marx (1967), ibid.

[52]Cf. Sartre (1943).

[53]"*... the world is the will to power—and nothing besides!*" Cf. Nietzsche (1968, §1067).

(at least partly) from earthly burdens like diseases, hard physical or boring mental word, poorness, self-inflicted immaturity and starvation, such promises are welcome and seductive. Thus, the development costs for such technologies must be paid in advance, and this is done by public and private grants and third-party funds and by selling products, promising AI and containing semi-smart programs for autonomous cars, healthcare robots, smart homes, etc. The need of AI research is substantiated by the alleged immediate benefit of technology in health care due to an ageing society, production and service, mobility, logistics, advertisement, governance, security and communication. Nevertheless, it is not alone the interest of those people, who are earning money and living from AI research. Moreover, the producers of computer and network hardware are also interested in new perspectives, which force the prospective clients to purchase new devices and equipment and which make them dependent on services and support. Moreover, the number of the human workforce can be reduced considerably by AI. This is an actual goal.[54]

A more far-reaching interest can be figured out when observing the monopolistic tendencies of the companies propagating AI as an inevitable way to singularity: With a monopolistic position in developing algorithms, programs and organizational methods of data harvest as well as selling the corresponding hardware, such companies could be enabled to master organizational forms of governance, of social structuring and—above all—of political agenda setting.[55]

5.3 Epistemological Interests

Fair enough we have to admit certain epistemological interest in AI. It is a legitimate purpose, to test hypotheses about cognition processes in man by observations, experiments and simulation. Moreover, an engineering basic principle states: Only what you have built, you can understand. Thus, it is tried to imitate, simulate and even to generate cognitive processes by machines whose principles are known. This motivation for basic research has led to a convergence of neuroscience, cognition research, AI and psychology with worldwide impressive budgets.[56] Nevertheless, there are many doubts about whether it will be possible, to explain cognitive processes by simulating them on Turing-like or von Neumann-type machines.[57]

[54]Roose (2019). The studies range from 40% up to 100% in the next 10 years.

[55]Ferenstein (2017) and Thiel and Masters (2014).

[56]For example the Brain Project, https://www.humanbrainproject.eu/en. For European citizens' concern about dual use of this research, see Badum and Jørgensen (2018). Explaining by making is a kind of technological reductionism. Proponents are e.g. Steinbuch (1961, 1965), Turing (1950), Simon (1970, 1979), and Goertzel and Pennachin (2007).

[57]Von Neumann (1951), Penrose (1991), Diamant (2015), Dreyfus (1992), Hermes (1971), Kornwachs (1989), Searle (1980), Shai et al. (2019), Toffoli (1982), Winograd (1990), and Winograd and Flores (1986).

6 A New Man, a New World: For Whom?

6.1 Religious Borderlines

The genius Alan Turing who had decoded the Enigma deciphering machine of "Deutsche Wehrmacht" developed some presentiments that machines could take control over the planet. The first idea of transhumanism was coined by Stanislav Ulam, the co-designer (together with Edward Teller) of the hydrogen bomb.[58] Much later, the physicist Stephen Hawking expressed the fear that the development of a matured artificial intelligence would mean the end of mankind.[59]

Nevertheless, we have to discuss the matter that there are engineers and computer scientists whose aim is indeed to make machines more intelligent than human beings, and which are enabled to start their own evolution. Will this be the paradise? A world is announced allowing a careless life without efforts, and with the possibility to upload our minds onto immortal, non-biologic machine basis. The machines will be better, smarter and much wiser than we,[60] and therefore, they will develop and improve themselves faster than human evolution, cultural and technological human enhancement. The paradise is paid with the superiority of machines over the man; they will know better what may be good for individuals, and for society. Will they become gracious bullies? According to different variants of this prophecy, there will be either an evolutionary melting or convergence between man and machine to a new man[61] or the distinction of the human species.[62]

The man has always tried to gain relief by techniques and technologies with the aim to live longer, without exertion, and with more possibilities. Nevertheless, this relief has nothing to do with redemption within the context of nearby all religions—let it be a post-mortal life, free from burdens and debts of earthly existence, let it be a dissolution in non-existence, be it a transcendence process for man, becoming a new man and a new earth.

Here, we can find a quasi-religious version of the redemption thought: The machine surmounts the frailness of human being, making him a new man, either by symbiosis with machines or by substituting him and taking his role in the world. Due to a naturalistic view of man, any interaction between machine and man is considered at least based on information flows as physical interactions. Insofar the man–machine interaction will transform the man into another sort of man–machine entity; this change is based on technical and biological, i.e. material, processes. In

---

[58]See Bostrom (2005), with reference to Teller (2003) and Ulam (1958, p. 5).

[59]Cf. Cellan-Jones (2014). There is also a manifesto of scientists, warning against the development and use of autonomous, intelligent weapons; cf. Autonomous weapons (2016).

[60]"*Computers share knowledge much more easily than humans do, and they keep that knowledge longer, becoming wiser than humans*". Lisi (2015).

[61]Bostrom (2007).

[62]Hawking et al. (2014) and Harari (2017).

this view, material processes are described only by physics. The interaction between information and physical systems can be conceived as processes in computers.

Behind this idea we can find a computer-centred concept of nature: Physics is something like software. Molecules and atoms behave only seemingly according to natural laws; rather they actually execute what has been calculated, e.g. the angles between atoms with respect to chemical bonding. Physics becomes a simulation of physics. Kevin Kelly expressed this slightly ironically:

> An ultimate simulation needs an ultimate computer, and the new science of digitalism says that the universe itself is the ultimate computer—actually the only computer. Further, it says, all the computation of the human world, especially our puny little PCs, merely piggybacks on cycles of the great computer. Weaving together the esoteric teachings of quantum physics with the latest theories in computer science, pioneering digital thinkers are outlining a way of understanding all of physics as a form of computation.
>
> From this perspective, computation seems almost a theological process. It takes as its fodder the primeval choice between yes or no, the fundamental state of 1 or 0. After stripping away all externalities, all material embellishments, what remains is the purest state of existence: here/not here. Am/not am. In the Old Testament, when Moses asks the Creator, "Who are you?" the being says, in effect, "Am." One bit. One almighty bit. Yes. One. Exist. It is the simplest statement possible.[63]

The revelation to such a "religion" is based on two dogmas:

1. Computation is able to describe all existing things and processes.[64]
2. Everything that exists is a computer.[65]

Against this background, the ideology of "digitalism" promises to free us from the innate deficiency of earthly humanity, i.e. from the limited cognitive and intellectual abilities, the slowness, the disturbing emotions and most of all from our ageing biological system, say our body. This is to be done simply by becoming cyborgs or uploading our consciousness, represented in software to data storage.

A kind of church has even been founded: Since 2015, the organization "Way of the Future" founded by Anthony Lewandowski strives to develop and promote a deity based on artificial intelligence.[66] Behind all this, it seems to be the belief: If we increase our computing power more and more, we can achieve more and more unity with the one Big Computer.

---

[63]Cf. Kelly K. (2012). The reference is Exodus, 3:14.

[64]A forerunner of this idea is Archibald Wheeler (cf. Kelly K. 2012 and Carl F. von Weizsäcker 1980, chap. II, 5.4, using a dissertation of M. Drieschner). He proposed a quantum mechanics that can be reduced to a combination of simple twofold alternatives. The measuring process, i.e. the generalized concept of interaction at all, produces at least 1 bit, deciding a simple alternative. To be fair, this is not connected to any religious ideas, but to the trial, to find a simple, but universal formalism for quantum mechanics interaction. Cf. also Lloyd (2006).

[65]This concept has some forerunners, such as Zuse (1969, 1982).

[66]Cf. Harari (2017).

It is easy to draw even more parallels to religious moments[67]— it may be sufficient to say that the scientific and technical press has not concealed the missionary and partly fanatical features of this movement. Thus, the transhumanism based on singularity is not a methodologically criticizable science, but a kind of theology whose salvation prospects are not rosy to those who do not share this sect-like "religion". They have to stay outside of the digital paradise. That is why "digitalism" is a kind of degenerated theology, nothing more.

It is not said that transhumanism is necessarily connected with this ideology of digitalism, but certain kind of ideas can be found frequently by some protagonists.[68]

6.2 *Reductionism*

The singularity, particularly in the context of the digitalism discussed above, has some striking similarities with the chiliasm mentioned above. Both variants (1) and (2) can be found: Machines may lead either to an apocalypse or to a new brave world with all characteristics of a paradise, and the self-conscious man can upload his individual personality as a future existence in silico with no pain, burdens and debts.

The concept of the man behind this projection is necessarily and unavoidable reductionist[69]:

- Thoughts and contents of self-consciousness are only based on material processes (organic material), which can be described by algorithms. Within the frame of the concrete situation, initial and boundary conditions, they can be interpreted as identical with thoughts and contents of self-consciousness.
- *Man is a meat machine.*[70] A machine is defined by algorithms and physical processors performing the algorithm in terms of programs.
- The existence of man is defined by the conditions of the possibility of physical processes performing algorithms. It is not astonishing that this definition is co-extensive with the conditions of the possibility of the existence of computer or robots.
- Consciousness and self-consciousness are bounded to physicality and the self-sensory properties of the physical body.[71]
- Man and machine can learn from outer experiences in the same way—nevertheless, due to the difference of situations, constraints and initial conditions, the

[67]Braga and Logan (2018) have collected statements of proponents which show striking parallels to the Christian idea of salvation and eternal life.

[68]Ibid.

[69]There are some variants of reductionism. We refer here to an ontological reductionism; cf. Lucadou and Kornwachs (1983).

[70]Minsky, cit. Weizenbaum (1977, pp. 72–73).

[71]Kornwachs (2002, 2006).

result of learning is different even with same stimuli and cannot be forecast. This gives rise to an individuality not only man but also machines may develop.

Due to this individuality, and additionally, due to the ability to produce better machines as a man can do, machines will evolve from one production cycle to another. This process will be unstoppable under the conditions that uncontrolled (not supervised) learning is allowed to the machines and that the machines are enabled to build new machines in ExL modus.

Transhumanism looks forward to the end of this development: Machines with self-consciousness, able to set their own goals and aims, smarter than man, with a necessary ultimate goal of self-preservation will either strive to converge with man as an enhanced form of cyborgs or will strive to remove man as an annoying competitor for resources, energy and space.

Such projections are only possible under a strong reductionist concept of mind and man. The motivations to develop such theology like anthropologies may have to do with the rejection of the fourth insult, mentioned above. One strategy is to justify technological support of life by proofing its potentiality for reliefs of all kinds. In a secularized culture like in Western industrial societies, it is seductive to equalize relief with salvation. Moreover, the proof of feasibility protects from the accusation of hubris. Finally, the substitution of present Abrahamic religions Jewish, Christian, Islamic) by the religion of digitalism or variants thereof is connected with the desire to reconcile science, technological experience and religious longings.

7 Resist the Beginnings

7.1 *Objections*

The principal limits of AI have been discussed widely.[72] This discussion is closely connected with criticism of research programs and with exaggerated expectations, with the projections of future applications as well as with rejection due to ethical reasons. This connection had led to the dictum of Herbert Simon that there would be a logical contradiction to warn about the dangers of future AI application on the one hand and on the other hand to forecast that the aims of AI cannot be achieved in principle.[73]

The objections can be classified according to the following points:

[72]Braga and Logan (2018) and Weizenbaum (1977) did not believe that the examples for the methods of investigating human individuals are not suitable to generalize them to formally describable processes; and behaviour in a psychological laboratory would not be comparable with the complexity of everyday behaviour.

[73]H. A. Simon, talk at the University of Stuttgart. Oral Communication. For the controversy see Simon (1979), based on (1970) and Weizenbaum (1977). Simon believed that it would be possible to generalize experimental evidence about human cognitive behaviour to procedures and algorithm.

1. AI programs as far as they are implemented on a computer of von Neumann-type as well as emulated neuronal nets are restricted to the well-known limits of computability. Thus, there is no self-referential expression accepted, and it is not possible to compute real GOTOs.[74] In other words: A program on a Turing machine can alter itself only within the realm of finite and accountable alternatives. This enables machines to simulate learning, but genuine new alternatives cannot be produced, or if any, by statistical means. Moreover, it cannot be decided whether they are meaningful or not.[75]
2. Dedicated neuronal nets (i.e. hardware based, not emulated on Turing-type machines or combinations thereof) can execute processes which cannot be characterized anymore as algorithms in terms of computability.[76] Whereas there are considerable difficulties to prove the correctness of an arbitrary program on a Turing machine,[77] it is not possible to predict processes in real neuronal nets or to reconstruct them completely. Thus, we have no reproducible results with respect to input-path-output schemes.[78]
3. A reductionist AI concept of human cognition cannot take into account "*the full dimension of human intelligence*". These issues are characterized by "*curiosity, imagination, intuition, emotions, passion, desires, pleasure, aesthetics, joy, purpose, objectives, goals, télos, values, morality, experience, wisdom, judgment, and even humor.*"[79]
4. The actual and the prospective energy consumption efficacy of computers is still far away from biological possibilities: The Watson Computer having beaten a human Jeopardy player needed 200 kW; a brain needs ca. 20 W.[80]

Nevertheless, today we can no longer argue with the term hubris—either morally or philosophically. Because we do not even know whether such better systems can be built and programmed one day, and if any, whether they will take us the book out of our hands.

Scepticism may be appropriate, and it helps to keep in mind the distinction between structural and functional analogies. The Go computer program that defeated a human player *behaves* like an intelligent player, but whether it *is* already an intelligent player is worthy of questioning. If a robot seems to act morally, you cannot yet conclude that it is freely and autonomously acting. From behavioural similarities, one cannot conclude on the generating structure inside, but if one knew the structures and the basic behaviour of the elements, one could infer the overall

---

[74]Kornwachs (1989).

[75]Ben-David et al. (2019)

[76]Li et al. (2019).

[77]Floyd R. (1967) and Clarke E. (1979).

[78]Milano et al. (2018).

[79]Braga and Logan (2018).

[80]See the energy consumption assessments by Sandberg (2015, 2016) and Wolpert (2018).

behaviour.[81] In the case of the computer, this is true; in humans, we probably do not know the structure of the psyche and the brain well enough. Therefore, we should not call a behaviour intelligent if we cannot even properly state what human intelligence is, where it comes from and by what it has been caused. Therefore, the temptation of AI's protagonists is great, to define intelligence exactly according to what their machines already can do.

7.2 *Moralizing Business Models*

Rather we should ask the question, if and if so, why and to what extent we want to have such autonomized systems at all. The advantages of driverless mobility, care robots, mine-detection systems, production equipment and perhaps even automatic weapons may be mentioned and discussed, but the moral dilemmas that the use of such systems will produce have already drawn serious attention.[82] What price are we willing to pay for the machines to do and manage everything for us? At this point, it is not yet about metaphysical compensation considerations that everything in life has its price, but about illuminating the path we are about to embark on and which is propagated by the propositions about transhumanism.

The proposals for this path come from people whose technological expertise is undisputed, but whose image of people and society horrifies those not sharing the same enthusiasm homed in Silicon Valley. To believe that this movement would be the attempt to improve the world fails to recognize that behind all propagated technologies and future images, painted with bright colours, there are interests in the form of business models. Anyone who thinks that Google and Co want to bring us benefits has not understood that yet.[83] It is, therefore, time, not only to subject the planned technologies to an ethical evaluation and technology assessment—which is already done but too little intensive—but also to undergo the underlying business models to a moral and impact assessment. Sometimes, they seem to transport a certain vision of world domination. An outraged debate on data protection is not enough.

Before we think of transhumanism as a future omega point of development, we should think of the humanism of the present time, which many people consider as endangered.[84] We should, therefore, define some limits for the research and design

[81]Under certain circumstances, e.g. for electrical circuits when the basic behaviour of the elements is linear in the first approach. See Padulo and Arbib (1974) and textbooks in electronics.

[82]It is not the question whether robots may be moral agents (Loh 2019), but the moral dilemmas with which the users and concerns are confronted; cf. Müller (2016). Thus, AI can play the role of an error amplifier: The learning results are biased if the data are biased. The data are biased when the collector of data has a biased theory.

[83]Cf. *pars pro toto* a discussion by Spehr and Weber (2016) about smart home. A survey of new business models and value chain is given in Jung and Kraft (2017).

[84]Cf. Harari (2017); for a modern concept of humanism, see also Fromm (2005).

of artificial intelligence products. The field of robot ethics is beginning to think about it, but it is still a young field.[85] We should not try to understand self-consciousness by building machines with simulated properties of self-consciousness before we do not actually know what human self-consciousness is.

Rather, with respect to current and future products, we should ensure that the difference between the human subject and the virtual agent remains visible to all individuals concerned. No machine should ever try to win the Turing test in an expanded form against the will of the users or any person concerned by their functionality. The machine must be recognizable as a machine; otherwise, we can potentially not say *no*, or technically speaking: to switch off.

There is an analogy to advertising—it must be always recognizable as an advertisement; otherwise, there is a monitum by a board. Perhaps, one can also consent to the demand of sociologist Harald Welzer: As long as dilemmas occur, as we discuss them with autonomous driving cars (e.g. Trolley problem[86]), one should not simply put such systems into operation.[87]

Is that too easy? Yes and no. Yes, because the temptations that the consumer and economic benefits of such a development are so promising that they could sweep away concerns of this kind. And no, because it belongs to the conditions of responsible action,[88] that one dominates the instrument with which one acts. Thus, we should resist the beginnings: If the instrument controls us, we are no longer free and can no longer take any responsibility.

7.3 Eight Rules

That is why man has to take his new creatures by the hand—with a firm grip. For the sake of application, eight rules for using machines can be formulated:

1. Never use a decision-making system that substitutes your own decision. Even robots must not be used in decision-making intent.
2. *Nihil Nocere*—do not tolerate any harm to users.
3. User rights break producer rights.
4. Do not build pseudo-autonomous systems that cannot be turned off. Fully autonomous systems should not be allowed.
5. The production of self-conscious, autonomously acting robots (if possible) is prohibited (analogous to the chimaera ban and human cloning ban in genetic engineering).

[85] Loh (2019) and Tzafestas (2016).

[86] Originally coined by Foot (1967).

[87] Cf. Welzer (2017).

[88] Kornwachs (2000).

6. Do not fake a machine as a human subject. A machine must remain machine; imitation and simulation must be always recognizable. It must always be clear to all people involved in human–machine communication that a machine communication partner is a machine.
7. If you do not know the question and the purpose of the question, you cannot handle the system response and understand the behaviour of a robot. The context must always be communicated.
8. Anyone who invents, who produces, operates or disposes of technology has interests. These interests must be disclosed honestly.

8 Concluding Remarks

The discussion, whether singularity will be possible,[89] whether the singularity is a presupposition and whether transhumanism will be near or not, remains on the level of forecasts, or better, prophecy. What we can do is to discuss possible consequences and to pose questions, in what kind of future we want to live, which functions of our life should be supported by AI and which not and with which values and their priorities we want to argue. The inevitable task to shape future technologies urges us to discuss our images about man and society. This includes arguments of philosophical anthropology. The analysed anthropology of transhumanism seems to be derailed; it is just bad theology and its position is still ahead of the Enlightenment.

References

Autonomous Weapons: An open letter from AI robotics researchers. https://futureoflife.org/open-letter-autonomous-weapons/ (2016)

Badum, N.B., Jørgensen, M.L.: European citizens' view on neuroscience and dual use. Synthesis report of citizen workshops. Human Brain Project—Ethics & Society; Danish Board of Technological Foundation. http://hbp.tekno.dk/wp-content/uploads/2018/03/Synthesis-Report-of-Citizen-Workshops.pdf (2018)

Ben-David, S., Hrubeš, P., Moran, S., Shpilka, A., Yehudayoff, A.: Learnability can be undecidable. Nat. Mach. Intell. **1**, 44–48 (2019). https://doi.org/10.1038/s42256-018-0002-3

Boden, M.: Artificial Intelligence and Natural Man. Harvester Press, London (1977)

Bostrom, N.: A history of transhumanism thought. J. Evol. Technol. **14**, 1–25 (2005)

Bostrom, N.: In defense of posthumanism dignity. Bioethics. **19**(3), 202–2014 (2007)

Botsman, R.: Big data meets big brother as China moves to rate its citizens—the Chinese government plans to launch its social credit system in 2020. http://www.wired.co.uk/article/chinese-government-social-credit-score-privacy-invasion vom 21.10.2017 (2017)

Braga, A., Logan, K.R.: The emperor of strong AI has no clothes: limits to artificial intelligence. Information. **8**, 156 (2018). https://doi.org/10.3390/info8040156

[89]Yampolsk (2018).

Cellan-Jones, R.: Stephen Hawking warns artificial intelligence could end mankind. BBC News, Dec 2. www.bbc.com/news/technology-30290540 (2014)

Clarke, E.: Programming language constructs for which it is impossible to obtain good Hoare-like axiom systems. J. Assoc. Comput. Mach. **26**(1), 129–147 (1979). https://doi.org/10.1145/512950.512952

Dawkins, R.: The Extended Phenotype. University Press, Oxford (1982)

de Chardin, T.: L'Avenir de l'Homme. Eng. (2002): The Future of Man. Image Books, New York (1964)

Dennett, D.C.: The Intentional Stance. MIT Press, Bradford, Cambridge (1987)

Diamant, E.: Advances in artificial intelligence: are you sure, we are on the right track? http://arxiv.org/abs/1502.04791; http://arxiv.org/ftp/arxiv/papers/1502/1502.04791.pdf (2015)

Dreyfus, H.L.: What Computers Still can't Do. A Critique of Artificial Reason. MIT Press, New York (1992)

Ferenstein, G.: The disrupters—silicon valley elites' vision of the future. City J. Winter 2017. https://www.city-journal.org/html/disrupters-14950.html (2017)

Floyd, R.W.: Assigning meanings to programs. Proc. Am. Math. Soc. Symp. Appl. Math. **19**, 19–31 (1967)

Foot, P.: The problem of abortion and the doctrine of double effect. Oxford Rev. **5**, 5–15 (1967)

Fromm, E.: Humanismus als reale Utopie. Ullstein, Berlin (2005)

Fuller, S., Lipinska, V.: The Proactionary Imperative: A Foundation for Transhumanism. Palgrave Macmillan, London (2014)

Gehlen, A.: Die Seele im technischen Weltalter—sozialpsychologische Probleme in der industriellen Gesellschaft. Rowohlt, Reinbeck (1957)

Goertzel, B., Pennachin, C. (eds.): Artificial General Intelligence. Springer, New York (2007)

Grunwald, A.: Converging Technologies for human enhancement—a new wave increasing the contingency of the condition humana. In: Banse, G., Grunwald, A., Hronszky, I., Nelson, G. (eds.) Assessing Societal Implications of Converging Technological Development. Proceedings of the 3rd Workshop for Converging Technologies, Budapest, Ungarn, 8–10 Dec 2005, pp. 271–288. Edition Sigma, Berlin (2007)

Harari, Y.N.: Homo Deus—a history of tomorrow. Homo Deus. Harvill Secker. Penguin, London (2017)

Hawking, S., Russell, S., Tegmark, M., Wilczek, F.: Stephen Hawking: 'Transcendence looks at the implications of artificial intelligence—but are we taking AI seriously enough?' The Independent 05–01, Thursday May 1. https://www.independent.co.uk/news/science/stephen-hawking-transcendence-looks-at-the-implications-of-artificial-intelligence-but-are-we-taking-9313474.html (2014)

Heidegger, M.: Sein und Zeit (1927). Niemeyer, Tübingen. Eng.: Being and Time. State University of New York Press, New York (1986)

Hermes, H.: Aufzählbarkeit, Entscheidbarkeit, Berechenbarkeit. Springer, Berlin (1971)

Heylighen, F., Bollen, J.: The World-Wide-Web as a super-brain: from metaphor to model. In: Trappl, R. (ed.) Cybernetics and Systems' 96, pp. 917–922. Austrian Society for Cybernetics. http://pespmc1.vub.ac.be/Papers/WWW-Super-Brain.pdf (1996)

Heylighen, F., Lenartowicz, M.: The global brain as a model of the future information society: an introduction to the special issue. Technol. Forecast. Soc. Chang. **114**, 1–6 (2017)

Husserl, E.: Logische Untersuchungen 1900 Eng.: Logical Investigations. Routledge & Keagan, London (1970)

Joy, B.: Warum uns die Zukunft nicht braucht. Frankfurter Allgemeine Zeitung Nr. 130, Jun 6, pp. 49–51. Eng.: The Future doesn't need us. WIRED 8.04. www.wired.com/2000/04/joy-2/ (2000)

Jung, H.H., Kraft, P.: Digital vernetzt—Transformation der Wertschöpfung. Szenarien, Optionen und Erfolgsmodelle für smarte Geschäftsmodelle, Produkte und Services. Hanser, München (2017)

Kant, I.: Kritik der Reinen Vernunft (1781). In: Werkausgabe, hrsg. von W. Weischedel, Bd. III–IV. Suhrkamp, Frankfurt am Main (1995)

Kant, I.: Logik (1791). In: Kant, I. Werkausgabe, hrsg. von W. Weischedel, Bd. VI. Suhrkamp, Frankfurt am Main (1995)
Kelly, K.: God is the machine. WIRED, Jan 12. https://www.wired.com/2002/12/holytech/ (2012)
Kelly, S.D.: A philosopher argues that an AI can't be an artist. MIT Technology Review, Feb 21 (2019)
Kornwachs, K.: Self reference and information. In: Dalenoort, G. (ed.) The Paradigm of Self-Organization, pp. 309–321. Gordon & Breach, London (1989)
Kornwachs, K.: Das Prinzip der Bedingungserhaltung. Eine ethische Studie. Reihe Technikphilosophie Bd.1. Lit. Münster, London (2000)
Kornwachs, K.: Bewusstsein, Programm, Körper. In: Kegler, K.R., Kerner, M. (eds.) Der Künstliche Mensch—Körper und Intelligenz im Zeitalter ihrer technischen Reproduzierbarkeit. Köln, Böhlau (2002)
Kornwachs, K.: Knowledge + Skills + x. In: Dengel, M., Roth-Berghofer, T., et al. (eds.) Wissensmanagement 2005. Postproceedings der Wissensmanagement Konferenz 2005. Lecture Notes in Computer Science, pp. 32–47. Springer, Heidelberg (2006)
Kornwachs, K.: Arbeit 4.0—People Analytics—Führungsinformationssysteme: Soziologische, psychologische, wissenschaftsphilosophisch—ethische Überlegungen zum Einsatz von Big Data in Personalmanagement und Personalführung. Gutachten für die Universität Münster, Vergabenummer 2017_59_BS. Büro für Kultur und Technik, Argenbühl-Eglofs, 28 Feb (2018)
Kurzweil, R.: The Age of Spiritual Machines: when Computers Exceed Human Intelligence. Penguin Books, New York (1999)
Kurzweil, R.: The Singularity Is near. Penguin Books, New York (2006)
Landmann, M.: Philosophische Anthropologie. de Gruyter, Berlin, New York (1982)
Leibniz, G.W.: Vernunftprinzipen der Natur und der Gnade. Meiner, Hamburg (1982)
Levit, G.S.: The biosphere and the noosphere theories of V. I. Vernadsky and P. Teilhard de Chardin: A Methodological Essay. Int. Arch. Hist. Sci./Arch. Int. Hist. Sci. **50**(144), 160–176 (2000)
Li, C., Wang, Z., Rao, M., Belkin, D., Song, W., Jiang, H., Yan, P., Li, Y., Lin, P., Hu, M., Ge, N., Paul Strachan, J., Barnell, M., Wu, Q., Williams, R.S., Yang, J.J., Qiangfei, X.: Long short-term memory networks in memristor crossbar arrays. Nat. Mach. Intell. **1**(1), 49–57 (2019). https://doi.org/10.1038/s42256-018-0001-4
Libet, B.: Do we have a free will? J. Conscious. Stud. **5**, 49 (1999)
Lisi, A.G.: I, for one, welcome our machine overlords. In: Brockman, J. (ed.) What to Think About Machines that Think, pp. 22–24. Harper Perennial, New York (2015)
Lloyd, S.: Programming the Universe. Knopf Doubleday, New York (2006)
Logan, R.K.: Can computers become conscious, an essential condition for singularity? Information. **8**, 161 (2017). https://doi.org/10.3390/info8040161
Loh, J.: Roboterethik—eine Einführung. Suhrkamp, Frankfurt am Main (2019)
Lovelock, J.: Gaia: the Practical Science of Planetary Medicine. Gaia Books, London (1998)
Lunceford, B.: Love, emotion and the singularity. Information. **9**, 221 (2018). https://doi.org/10.3390/info9090221
Lyotard, F.: The Postmodern Condition: A Report on Knowledge. University of Minnesota Press, Minneapolis (French: La Condition postmoderne: Rapport sur le savoir. Éditions de Minuit, Paris, 1979) (1984)
Lyotard, J.-F.: The Differend: Phrases in Dispute. University of Minnesota Press, Minneapolis (1988)
Madelung, W.: al-Mahdi. In: Beraman, P., et al. (eds.) Encyclopedia of Islam, vol. 5, 2nd edn. Leiden, Brill (1986)
Margulis, L.: Symbiotic Planet: a New Look at Evolution. Basic Books, New York (1999)
Marx, K.: Das Kapital, Bd. 1. In: Marx Engels Werke (MEW) Bd. 23. Dietz, Berlin (Marx, K.: Capital, Volume I, Part III, Chap. 7, Sec. 1. Transl. by Ben Fowkes. Penguin Books, London (1990) [1867]). https://www.marxists.org/archive/marx/works/1867-c1/ch07.htm (1967)

Mewes, F.: Regulations for the Human Park: On Peter Sloterdijk's Regeln für den Menschenpark. Gnosis VI, p. 1 (2002)
Milano, G., Luebben, M., Ma, Z., Dunin-Borkowski, R., Boarino, L., Pirri, C.F., Waser, R., Ricciardi, C., Valov, I.: Self-limited single nanowire systems combining all-in-one memristive and neuromorphic functionalities. Nat. Commun. **9**, 5151 (2018). https://doi.org/10.1038/s41467-018-07330-7
Misselhorn, C.: Grundfragen der Maschinenethik. Reclam, Stuttgart (2018)
Moravec, H.: Mind Children—The Future of Robot and Human Intelligence. Harvard Press, Cambridge (1990)
More, T.: Utopia (in Latin). Eng.: Utopia (Dolan, J.P., trans.). In: Green, J., Dolan, J.P. (eds.) The Essential Thomas More. New American Library, New York (1516/1967)
More, M.: The Extropian Principles, p. 1, Jul 25. Revised Jun 1994. www.aleph.de/Trans/Cultural/Philosophy/princip.html (1993)
Müller, V.C.: Risk of Artificial Intelligence. CRC Press—Chapmann & Halls, London (2016)
Nietzsche, F.: In: Kaufmann, W. (ed.) The Will to Power (Literary Remains, Posthum). Vintage, Vancouver (1968)
Padulo, L., Arbib, M.A.: System Theory—Unified State-Space Approach to Continuous and Discrete Systems. Hemisphere, Toronto (1974)
Penrose, R.: Computerdenken. Spektrum, Heidelberg. Eng.: The Emperors new Mind. Oxford University Press, Oxford (GB) (1991)
Pfleger, K.: Die verwegenen Christozentriker—Augustinus, Pascal, Luther, Schelling, Solowjew, Teilhard de Chardin. Herder, Freiburg (1964)
Pico della Mirandola, G.: Oratio de hominibus dignitate. Engl. Translation: eBooks@Adelaide, The University of Adelaide Library, South Australia 5005. https://ebooks.adelaide.edu.au/p/pico_della_mirandola/giovanni/dignity/. For Latin text see: https://la.wikisource.org/wiki/Oratio_de_hominis_dignitate (1496)
Popper, K.: Die offene Gesellschaft und ihre Feinde, 2 Bde. Mohr, Siebeck, Tübingen. Eng. (1945): The Open Society and Its Enemies. Routledge, London (1992)
Ratzinger, J. (later Pope Benedictus XVI): The Spirit of the Liturgy. Ignatius Press, Kindle Edition, San Francisco, Nov 6 (2006)
Roco, M.C., Bainbridge, S. (eds.): Converging Technologies for Improving Human Performance. Nanotechnology, biotechnology, information technology, and cognitive science. National Science Foundation (NSF/DOC) sponsored report. Arlington, Virginia, January (2002)
Roco, M.C., Bainbridge, W.S., Tonn, B., Whitesides, G. (eds.): Convergence of Knowledge, Technology and Society—beyond Convergence of Cognitive Nano-Bio-Info-Cognitive Technologies. Springer Swizerland, Cham (2013)
Roose, K.: The hidden automation agenda of the Davos elite. New York Times, Jan 25 (2019)
Roth, G., Strüber, N.: Wie das Gehirn die Seele macht, 7th edn. Klett-Cotta, Stuttgart (2017)
Sandberg, A. Energy requirements of the singularity. Andart II, Feb 2. http://aleph.se/andart2/megascale/energy-requirements-of-the-singularity/ (2015)
Sandberg, A.: Energetics of the Brain and Ai. Sapience Project 2016. Technical Report STR 2016, London Feb 2. https://arxiv.org/pdf/1602.04019v1.pdf (2016)
Sartre, J.-P.: L' Être et le Néant. Gallimard, Paris. Eng.: Being and Nothingness. Routledge, London (1943)
Searle, J.: Minds, Brains and Programs. Behav. Brain. Sci. **3**(3), 417–457 (1980). https://doi.org/10.1017/S0140525X00005756
Shai, B.D., Hrubes, P., Moran, S., Shpilka, A., Yehudayoff, A.: Learnability can be undecidable. Nat. Mach. Intell. **44**(1), 44–48 (2019). www.nature.com/natmachintell. https://doi.org/10.1038/s42256-018-0002-3
Simon, H.A.: The Sciences of the Artificial. MIT Press, Cambridge, MA (1970)
Simon, H.A.: Models of Thought. Yale University Press, New Haven (1979)

Singer, W.: Synchrony, oscillations, and relational codes. In: Chalupa, L.M., Werner, J.S. (eds.) The Visual Neurosciences, pp. 1665–1681. The MIT Press, A Bradford Book, Cambridge, MA (2004a)
Singer, W.: Verschaltungen legen uns fest. Wir sollten aufhören, von Freiheit zu sprechen. In: Geyer, C. (Hrsg.) Hirnforschung und Willensfreiheit. Zur Deutung der neuesten Experimente, S. 30–65. Suhrkamp, Frankfurt am Main (2004b)
Sloterdijck, P.: Regeln für den Menschenpark. Ein Antwortschreiben zu Heideggers Brief über den Humanismus. Frankfurt am Main, Suhrkamp (1999)
Spehr, M., Weber, L.: Die smarte Kapitulation. Frankfurter Allgemeine Zeitung Nr. 3, Jan 5, S. T1 (2016)
Steinbuch, K.: Die Lernmatrix. Kybernetik. **1**(1), 36–45 (1961)
Steinbuch, K.: Automat und Mensch. Springer, Berlin (1965)
Teller, E.: Ich musste es tun. Interview mit J. Schönstein. FOCUS Magazin 38, Sept 15 (2003)
Thiel, P., Masters, B.: Zero to One—Notes on Start Ups or How to Build the Future. Virigin Books, Penguin, London (2014)
Tipler, F.J.: The Physics of Immortality. Doubleday, New York (1994)
Toffoli, T.: Physics and computation. Int. J. Theor. Phys. **21**(3/4), 165–175 (1982)
Trepl, L.: Allgemeine Ökologie. Peter Lang, Frankfurt am Main (2005)
Trepl, L.: Die Erde ist kein Lebewesen—Kritik der Gaia-Hypothese. https://scilogs.spektrum.de/landschaft-oekologie/die-erde-ist-kein-lebewesen-beitrag-zur-kritik-der-gaia-hypothese/ (2013)
Turing, A.M.: Computing machinery and intelligence. Mind. **49**, 433–460 (1950)
Tzafestas, S.G.: Roboethics—A Navigating Overview. Springer, Berlin (2016)
Ulam, S.: Tribute to John von Neumann. Bull. Am. Math. **64**, 1–49 (1958)
United State Department of Defense: Unmanned systems integrated roadmap. FY 2011–2035, Reference Number 11-S-3613. http://www.dtic.mil/dtic/tr/fulltext/u2/a558615.pdf (2011)
Vernadskij, V.I.: Der Mensch in der Biosphäre. Zur Naturgeschichte der Vernunft. Hrsg. v. Wolfgang Hofkirchner. Peter Lang, Frankfurt am Main (1997)
Vernadsky, V.I.: The Biosphere, in Russian (1926). Engl. Copernicus, New York (1997)
von Glasenapp, H.: Die fünf Weltreligionen. Diederichs, Berlin (1996)
von Lucadou, W., Kornwachs, K.: The problem of Reductionismfrom a system theoretic standpoint. In: Zeitschrift für Allgemeine Wissenschaftstheorie. **14**(2), 338–349 (1983)
von Neumann, J.: The general and logical theory of automata. In: Jeffress, L.A. (ed.) Cerebral Mechanisms in Behavior—the Hixon Symposium, pp. 1–14. Wiley, Oxford (GB) (1951)
Voss, P.: From narrow to general AI. Medium, Oct 3. https://medium.com/intuitionmachine/from-narrow-to-general-ai-e21b568155b9 (2017)
Weizenbaum, J.: Computer Power and Human Reason. From Judgement to Calculation. W. H. Freeman, San Francisco (1976)
Weizenbaum, J.: Die Macht der Computer und die Ohnmacht der Vernunft. Suhrkamp, Frankfurt am Main (1977)
Weizsäcker, C.F.: Die Einheit der Natur. Hanser, München. Eng. (1980): The Unity of Nature. Farrar, Straus & Giroux, New York (1971)
Welzer, H.: Schluss mit der Euphorie. DIE ZEIT 10, S. 6, Apr 27 (2017)
Winograd, T.: Thinking machines: can there be? Are we? In: Sheehan, J., Sosna, M. (eds.) The Boundaries of Humanity: Humans, Animals, Machines. University of California Press, Berkeley (1990)
Winograd, T., Flores, F.: Understanding Computers and Cognition: A New Foundation of Design. Ablex, New York (1986)
Wolpert, D.: Why do computers use so much energy? Scientific American, Oct 4. https://blogs.scientificamerican.com/observations/why-do-computers-use-so-much-energy/ (2018)
Yampolsk, R.V.: The singularity may be near. Information. **9**, 190 (2018). https://doi.org/10.3390/info9080190
Zuse, K.: Rechnender Raum, Schriften zur Datenverarbeitung Band 1. Friedrich Vieweg & Sohn, Braunschweig (1969)
Zuse, K.: The computing universe. Int. J. Theor. Phys. **21**(6/7), 589–600 (1982)

Chapter 3
Transhumanism and Philosophy of Technology

Guglielmo Papagni

Abstract At first glance, transhumanism may seem to be a proper framework for interpreting technological advancement, because of its positive attitude towards innovation, and especially when opposed to "technophobic" movements. Nevertheless, in the first part of this chapter it is demonstrated how the transhumanist premises are wrong, and how the movement itself represents, just as much as technophobic approaches, an extreme derivative of the Modern Western tradition. Specifically, the criticism is focused on the propensity of transhumanism to read the technological otherness from a utilitarian perspective, as a way to emancipate our species from the realm of nature. Instead, the aim of the second part is to offer a radical revision of the traditionally dichotomous relationship between "nature" and "culture", from which the transhumanist positions stem. In doing so, the chapter draws on different and more ontologically inclusive conceptual perspectives, namely the ones on which critical posthumanism is grounded. As a result, technology comes to play a new active role in the process of "shaping the human". Eventually, rather than being just a tool at our disposal, technology and technique are read as being a "species peculiar" pillar, and thus part of our very nature.

1 Introduction

When talking about the exponential growth that the "technosphere" is experiencing in our epoch, it is quite easy to see the great number of concerns that arise with it. Some of these preoccupations are reasonable; some others are, instead, rather naive. Among the latter, for example, we could put the so-called neo-Luddites movements which, as the name suggests (recalling the nineteenth-century movement), tend to fear the coming of intelligent machines that may eventually overtake their human creators. On the other hand, another kind of more optimistic interpretation is gaining popularity. It is what we could subsume under the label of

G. Papagni (✉)
Institute of Management Science, TU Wien, Vienna, Austria

W. Hofkirchner, H.-J. Kreowski (eds.), *Transhumanism: The Proper Guide to a Posthuman Condition or a Dangerous Idea?*, Cognitive Technologies,
https://doi.org/10.1007/978-3-030-56546-6_3

"transhumanist" movements. At first glance, and despite the reasonable concerns, this may seem to be a proper framework for addressing the topic of technological advancement, because of its positive attitude towards innovation.

Thus, the point I will try to make in this essay is not whether one of the two approaches is more appropriate than the other. I will shortly focus on the reasons why (keeping in mind what we already addressed as reasonable concerns) I do not think that standing against progress is a viable path. The main aim of this essay is rather to demonstrate, referring to transhumanism, why the premises themselves are wrong. Starting from this assumption, in the second part I will try to offer an alternative interpretation for addressing technology-related issues, though, not in the form of a Hegelian synthesis of the aforementioned approaches. I will rather refer to different and more ontologically inclusive conceptual perspectives, in order to give technology the place it deserves in the "making of the human".

To understand the meaning of the point I made, we have to look for the genesis of the concept of a discontinuity between nature and culture in the field of classical philosophy (as often happens). The reason for this lies in the fact that both the movements we introduced refer to our relationship with technology as being determined by an ontological separateness, which allows us to interpret it in an instrumental way. In these terms, as the Italian philosopher Roberto Marchesini has exposed, the split occurred at the very moment in which we started producing reflections on the human condition. The incompleteness paradigm—as Marchesini calls it—defines "the human" as a biologically incomplete species, when compared to the virtues with which other species have allegedly been endowed. He traces back the origin to the Promethean mythology. Basically, the myth handed down by Plato and Hesiod reports how, at the time of the distribution of the virtues among animal species, man would have been excluded from the "divine concession". This myth, Marchesini continues, treats human beings as being biologically inadequate in comparison with other animals. For this reason, the human "needs to drink in the cup of divinity thanks to Prometheus who gives (mankind) technical knowledge [entechnos Sophia] and fire" (Marchesini 2009, p. 18, translation of the author).

The Western culture has kept re-proposing this idea through the centuries. Paraphrasing a popular statement from Alfred North Whitehead, the European philosophical tradition fundamentally consists of a (long) series of footnotes to Plato (Whitehead 1978). An important moment of the process is represented by Cartesian formulations, and by the suggestion that, not adhering uniquely to the laws of matter (*res extensa*), humans have been able to rise above the animal/natural world. Although it is not among the purposes of this essay to show the evolution of this concept through the history of the Western culture, it can be useful to underline how it survived up to our times. Still in 1999, the Italian philosopher Umberto Galimberti was saying that "man is abysmally distant from the animal because it lacks the typical connotation of the animal which is instinct" (Galimberti 1999, p. 35, translation of the author). Because of this deficiency, "man, to live, is forced to build that complex of artifices, or techniques, able to make up for the insufficiency of those natural codes" (idem, p. 89). Notwithstanding the evolutionary continuity introduced by Darwin, and the more recent confirmations from neuroscience, the idea of this biological disadvantage, and therefore of a nature to be overtaken, has survived and crystallized in contemporary technophilic movements.

2 Technophobia and Technophilia

2.1 Technophobic Derivative

Concerning technophobic approaches, it would take us too far to address all the topics and feelings that could fit within a similar definition (as well as for so-called neo-Luddites movements) (Kiberd 2015). Some of the concerns related to technology are definitely reasonable and should be seriously taken into account. Thus, I do not mean to address explicitly the idea of a harmful exploitation of technological means. In fact, it seems quite obvious to state that, given certain technical knowledge (e.g. the capability of handling and controlling nuclear reactions), there are plenty of ways to apply it to comply with "wrong" purposes (e.g. nuking other nations). Some might say that, talking about atomic power, the threat of a nuclear fallout is precisely what prevented and kept the world safe from a (third) global war. But this kind of equilibrium based on the fear of extinction can hardly match with the idea of a peaceful and controlled technological development. Besides, I consider this idea as something we might generally long for; thus, this is not the extreme negative stance I mean to criticize. In fact, the perspective that I consider naive and inapplicable is the one that demonizes technology itself, aiming for a throwback to an allegedly glorious past condition. The point is that, even if it would make any sense to long for an alleged golden age of harmonious and respectful co-existing with nature (a sort of revival of Rousseau's noble savage), we can dismiss this hypothesis for its naivety and since it seems to ignore many of the pragmatic problems of such an idealization. As Ray Kurzweil has correctly noticed on his website, "how many people in the year 2000 would really want to go back to the short, disease-filled, poverty-stricken, disaster-prone lives that 99% of the human race struggled through a couple of centuries ago? We may romanticize the past, but until fairly recently, most of humanity lived extremely fragile lives, in which a single common misfortune could spell disaster" (Kurzweil 2001). The chances of reducing human suffering, says Kurzweil, represent an important reason for continuing to march on the path of technological development. The list of examples that may confirm this is quite long, and modern medicine (with the life span extension that it brought at least in the Western world) is just the most emblematic. Thus, we can easily dismiss these kinds of technophobic approaches by saying that it is not technology itself that represents a problem. It has never been up to technology to tell which of its application would have been harmful. Therefore, recalling what we just said about controlling atomic power, a more reasonable approach might consist in trying to drive positively the epochal transformations that characterize our time, instead of blindly opposing them. Through the thoughts of a self-declared transhumanist as Kurzweil, we can introduce the transition from the fears of technophobic accounts to the optimism towards technology of the transhumanist ones. For the sake of clarity, the choice of quoting one of the most influential transhumanists means that it is not among the intents of this essay to criticize technological progress (and optimism). Again, the main critique I aim to raise concerns the premises upon which transhumanism is growing.

2.2 *Transhumanism*

Following Marchesini, we can identify the transhumanist movement as a sort of reaction to the limits of humanism, as they have been enlightened by countercultural movements (Marchesini 2009, p. 521; the reference is to those branches of the Western philosophical tradition that, especially in the last century, developed radical critiques to humanism). Nevertheless, transhumanists put themselves in fundamental continuity with the leading tradition. In fact, as Max More, one of the leading thinkers of the movement and founder of the Extropy Institute, writes, "for several decades, it has been fashionable in some circles (especially the postmodernists and poststructuralists) to sneer at Enlightenment ideas, to declare that they are outdated, humancentric, or naive. Transhumanism continues to champion the core of the Enlightenment ideas and ideals—rationality and scientific method, individual rights, the possibility and desirability of progress [...]. In a philosophical rather than scientific form, Friedrich Nietzsche picked up this idea and declared that humans are something to be overcome, and asked 'What have you done to overcome him?' Although Nietzsche seemed not to see a role for technology in this transformation, his bold language inspired some modern transhumanists" (More 2013, p. 10). Here, we already see the expression of a sort of paradox, which is overcoming the human and its limits (as underlined by quoting one of the most influential anti-humanists as was Nietzsche), but leveraging on the battle horses of humanism itself, among which we see rationalism, individualism and anthropocentrism and from which an instrumental conception of technology, to the benefit of individual enhancement, is derived. Human enhancement, in such an individual conception (in terms of pursuing enhancement as a personal right), is a consequence of this approach and thus it expresses, as we previously noticed, the perpetration of ideologies grounded on a specific reading of the Promethean mythology. This, as we said previously, insists on the existence of a biological disadvantage that we can cope with thanks to the technological tool that our rationality has provided us with. In fact, states More, "Transhumanism is a class of philosophies of life that seek the continuation and acceleration of the evolution of intelligent life beyond its currently human form and human limitations by means of science and technology, guided by life-promoting principles and values" (More 2013, p. 1). This approach, underlines Domenico Parisi, seems to be successful (if for example compared to more critical posthumanist accounts), because our species is reticent to put itself in the broad picture that science is progressively unveiling, preferring instead to consider itself as something special and apart from the rest (Parisi 1999). In other words, it means that transhumanism tends to restore the traditional humanist hierarchy (in terms of relation between subject and object), within the framework of technocratic cyborgs. In practical terms, the transhumanist dream is that of emancipating the human from its mortal condition, letting it drift out from the biological evolutionary path to converge with machines. But, as Marchesini pinpoints, it is the subject who decides among the potential variety of technological solutions its way to get rid of the evolutionary burden. "Suffering, negative emotions such as fear, sadness, panic, depression, anxiety, boredom, irritability are seen as useless remains, the legacy of an evolutionary story to be abandoned in the name of a totally hedonistic and cognitive conception of being" (Marchesini 2009, p. 529, translation of the author).

In the work of authors such as Nick Bostrom or Max More, it becomes clear that what within humanism used to be the warrantor, i.e. faith in a deity, has been substituted by faith in technology. As Marchesini has underlined, "the inability to come to terms with humanism—that is, with the autarchic and separative idea of human ontology, the claim of a paradigmatic model of human ontology and the pervasive or projective notion of human ontology—and with anthropocentrism—that is, with the use of man as a measure of the universe, the vision of man as the end of the universe and the idea that human projectivity and expressiveness represent universal elements—has in fact caused serious damage to the development of an organic thought referred to the otherness and the relationship (also in the diachronic sense) with it; consequently, it is difficult to realize a proposal that can rightfully be declared to have overcome the self-referential paradigm" (idem, p. 512). In other words, on the one hand, the acceptation of the technological "infiltration" cannot coexist with the naive idea of humans being fully in control and capable of directing their own individual existence. On the other hand instead, it cannot fit with the notion, endorsed by authors such as Marvin Minsky, of a tyrannical perception of biology and nature (which is seen as full of shortcomings), which has, to say it with Nick Bostrom and Max More, to be turned into something at the disposal of the post-human being. They call it merging with technology (which, as I will try to show in the second part, might actually be the right approach), but in reality it is just exploiting technology from a subject-centred perspective. There cannot be merging if we do not include otherness (technological, in this case) into the ontological equation, and let it play an active role. This makes the transitional claim of transhumanism lose strength and meaningfulness because of the projective consideration towards technological beings. Whereas transhumanism tries to demonstrate its connections with the materiality of things, most of the times it actually perpetrates a dialectic of dualism, reproposing in new "cyborgian" forms the old dichotomies towards both nature and technique. Marchesini identifies this "post-Darwinian" overcoming, through "the development of an ever more invasive and pervasive technosphere" (idem, p. 533), as the biggest mistake of transhumanist accounts. By contrast, if we do not read technology as a standing against selective pressure but, from an "heteroreferential" stance, as a new selective partner, we become immediately aware that this hybridization with the technosphere does not result in a reduction in selection. It rather occasions a shift in the selective pressure or a change in the evolutionary context. Technique is inscribed in the flesh not only for the simple fact that we get used to considering it as embodied (as might be said of a single artefact), but above all because the technological partner, far from behaving inertly, becomes a new selective referent, our peculiar selective referent as a species. This means acknowledging the fact that technology is the peculiar expression of our species' equally peculiar genetic potential. Therefore, if we want to overcome the humanist limits of anthropocentric self-referentiality, it implies the need of integration of the natural and technological otherness. This will be a matter of analysis for the next part of this essay.

For a better understanding, we can recall a short study case as an example, for it exposes explicitly the mentioned limits. I am talking about the idea of "mind uploading" (more recently called "substrate independent mind", or SIM), a term coined back in 1988 by Hans Moravec in his book *Mind Children*. The reason why I chose to refer to this particular (hypothetical) practice is that it has become one of the marks and battle horses of transhumanism. The principle behind this concept is that, when machines will have reached a level of complexity similar to that of human beings in terms of "processing capability", it will be possible to transfer consciousness from the organic matter to another substrate capable of performing the same "operations".

2.3 *The Substrate-Independent Mind*

The main and most evident issue that can be identified in relation to the idea of a SIM is that a reduction in the direction of a functional dualism is needed to explain how it might work. "A functionalist holds that a particular mental state or cognitive system is independent of any specific physical instantiation, but must always be physically instantiated at any time in some physical form. [. . .] According to functionalism, mental states such as beliefs and desires consist of their causal role. That is, mental states are causal relations to other mental states, sensory inputs, and behavioral outputs. Because mental states are constituted by their functional role, they can be realized on multiple levels and manifested in many systems, including nonbiological systems, so long as that system performs the appropriate functions" (More 2013, p. 7). Whether they call it functionalism instead of dualism, it is quite clear that a reductionist process takes place in depicting the mind uploading process and, eventually, mind itself. The goal is that of "extracting" pure reasoning capability from mere physical matter (the parallelism with Cartesian *res cogitans* and *res extensa* is not by chance). This is, still pretty often, backed by the analogy between mind and computers, even though also authors following the functionalist approach have come to a recognition of the differential importance of the material substrate (Dennett 1995). Thus, the idea of reading human beings' cognitive capabilities as a matter of pure information independent from the substrate can still be seen as a direct consequence of a dualistic perspective. "Your brain is a material object. The behavior of material objects is described by the laws of physics. The laws of physics can be modeled on a computer. Therefore, the behavior of your brain can be modeled on a computer" (Merkle 2013, p. 157).

We see how a reductionist approach allows us to look both at biology as an obstacle and at technology as a means by which to cope with the obstacle. It is important to underline that the intent of this discussion is not to say that we will never achieve such a technological level which could allow similar performances. The critique is more on an epistemological and ontological level. Following Eddy Carli's argument, we cannot explain or describe human reason relying on the system of binary, digital logic ruling the computer world, even though this is often a tempting metaphorical approach (Carli 2000, p. 15). This position is backed by Domenico Parisi, who says that "perhaps it is the starting notion, the idea that the mind is a machine that manipulates symbols in an algorithmic way that is wrong [...]. Then comes the suspicion that it is the principle of ignoring the material basis of the mind that is wrong [...]. The software of a computer has not emerged over time from the hardware as the mind has emerged from the body" (1999, pp. 58–61, translation of the author). The idea of the emergence of biological forms of consciousness (or mind) throughout the billions of years of evolution of life sets a limit that is difficult to overcome relying on a functionalist and computational model. The contribution of neuroscience helps to enlighten the shortcomings of the latter approach. Gerald Edelman, for example, states that the nervous system's response configurations depend on the individual history of each system, which makes it deeply context dependent. In fact, it is only through interactions with the world that convenient response configurations are selected (Edelman 2000). Following Marchesini, we can identify the different nature between a computer, which manipulates symbols without purpose, and animal species which cannot be explained only synchronously because of the evolutionary process they are part of. Paraphrasing Antonio Damasio is the set of non-rational motivations that supports the tip of the iceberg of consciousness and reason, not the other way around (Marchesini 2009, p. 337). Most neurobiologists do not believe that the uploading of the mind is possible, for the very nature of the substrate itself, and of the ongoing processes (which are not algorithms) that we call mind. For both Gerald Edelman and Antonio Damasio, it is necessary to talk about cognitive performance as totally embodied functions and of the mind as an entity absolutely not separated from the biological dimension. According to Damasio, "body regulation, survival, and mind are intimately interwoven. The interweaving occurs in biological tissue and uses chemical and electrical signaling, all within Descartes' *res extensa*. Curiously, it happens most strongly not far from the pineal gland, inside which Descartes once sought to imprison the nonphysical soul" (Damasio 2005, p. 108). The ten to twelve billion synapses that compose the brain, together with the million billion connections that they generate, are therefore the critical point of this inextricable network, without considering the connection with the rest of the body. In fact, the diachronicity and context dependence of the neural activity are expressed by the fact that the latter is not at all self-sufficient. Following Marchesini, "astrocytes, particular glial cells, have for example the task of regulating the propagation of signals through the modification of the neuronal microenvironment, a complex repertoire of neuromodulators and hormones of endogenous or exogenous origin—for example taken with food—modify the performance of the synapses, the activated or inhibited areas of the brain, the growth of the ramifications of the dendrites and axons" (Marchesini 2009, p. 505, translation of the author). Thus, non-neural elements seem to play a leading role modifying how the brain works and influencing its structural genesis in a dynamic, i.e. diachronic, way. We can conclude this part saying that an enthusiastic and often uncritical approach towards technology seems to be misleading almost as much as a technophobic one, for it does not properly consider the importance of heteroreferentiality (of technological otherness and, as we have seen in the last case, of the flesh). The next question that arises is: can we refer to other theoretical stances within the Western cultural tradition, to address differently, more inclusively, the delicate and epochal topic of technological otherness?

2.4 *The Limits of the Humanist Heritage*

On the one hand, we have briefly seen how technophobic movements cannot provide a safe resort from uncontrolled technological development, at least for historical reasons (new technologies often pave the way for new possibilities), and for the naivety of the desire to return to alleged golden times of harmony between humans and nature. On the other hand, also transhumanism seems to entail different shortcomings. One is the paradoxical will to overcome our humanity, but pushing on the central features of humanism itself. Besides, another limit is that of not being capable of putting humans into the bigger picture enlightened by natural science, which has revealed how technology represents the peculiar expression of the genetic potential of our species. Finally, the (instrumental) enthusiasm towards powerful forms of technology seems to have two problems. We have seen (through the example of the SIM) the tendency to try to bend science to fit within a specific picture, as a symptom (second problem) of a frequently blind faith in technology. Somehow, this transhumanist belief in an almighty technology seems to restore the traditional humanist hierarchy, but within the framework of technocratic cyborgs. If this is correct, it is not more than giving old dichotomies a new shape.

If we switch our gaze for a moment in the direction of cross-cultural comparativism, we can gather hints on what kinds of alternative interpretations are available if we do not stick to the humanist paradigm. We can find an interesting (although brief) overview of such an alternative in an article published by *Wired* on the 30th of July 2018. Here, Joi Ito writes about the fundamental difference between the Western and the Japanese cultural backgrounds in terms of their approach to robots. Referring to the upcoming age of robotic technologies, he provides insights into the opposition between the generalized fear that is taking over the West, and the enthusiasm that the Japanese culture has been devoting to it over the last few decades. He identifies the main discriminant factor in the very difference of cultural—and religious—substrates upon which the two social systems are approaching this epochal change and the concept of the human condition itself. The point is that the two different ideas of humanity carry two different perspectives on the relationship that is assumed to exist between nature and culture. On the one hand, says Ito, we can see how "the Western concept of 'humanity' is limited" and leads to the belief that "we have the right to exploit the environment, animals, tools, or robots simply because we're human and they are not [. . .]. Followers of Shinto—continues Ito—unlike Judeo-Christian monotheists and the Greeks before them, do not believe that humans are particularly 'special'" (Ito 2018). With these words, the writer means that, from within the Shinto cultural system, we belong to nature more than it belongs to us, as a sort of spirit that permeates anything that exists, regardless of it being alive or not. I would like to take a cue from this example to show why we can trace back the mentioned limits of transhumanist positions to the same concept of humanity expressed by the Japanese author. In other words, this is not to say that the typical transhumanist fascination for technology is not a genuine enthusiasm for the possibilities displayed by modern technologies. The point I would like to make is more of an ontoepistemological kind, i.e. different premises might provide us with a better understanding.

3 Towards a Different Philosophy of Technology (and Nature)

In parallel to addressing the issue of technicity, some brief excursions in the realm of nature are needed, for these two topics are deeply and inextricably intertwined. Recalling the reading of the Promethean myth that I opted for in the beginning, we have seen that the source of the shortcomings of humanism is a wrong conception about the alleged biological disadvantage of our species. Within this framework, technology plays the instrumental role of a tool at the disposal of our mind (*res cogitans*) in its journey to unveil the world (*res extensa*). In the passage from classic humanism to its contemporary forms (i.e. transhumanism), the instrumental role did not really change, but its efficacy has been bolstered up to the point that technology has almost become the ultimate *telos* of the quest (i.e. the faith in technology that has taken the place of the traditional deity of classic humanism). The only element that stands stronger than technology itself is the "liberal individual" that Max More was referring to: that "Vitruvian Man" so strongly criticized by post-humanist thinkers, and by all those countercultural movements such as postcolonialism, feminism and so on. Thus, from modern transhumanist stances, technological knowledge has become the means by which to emancipate ourselves from our biological condition. The issues of nature and technicity are thus always deeply interwoven, up to the point that our technical skills themselves arise from nature, rather than being a countermeasure we erect to cope with alleged biological disadvantages. In the following paragraphs, I will try to develop these connections from different perspectives, to show how culture and technicity represent, indeed, our nature.

3.1 *Beyond the Failure of Modern Categories*

As we have seen at the beginning, within the modern tradition the technical object has been typically considered just a means by which our reason can extend its investigation to the outer world to cope with alleged biological limitations. Thus, if we would try to summarize the capital points of (trans)humanist stances, we could say that the Cartesian *res cogitans* still represents our meaningful essence, which is to be protected and preserved. The body is therefore something we might want to get rid of (as shown by the SIM example), and it only represents the annoying heritage of that natural burden that we are anxiously trying to leave behind us by means of an almighty technological capability. Concerning the Western philosophical tradition, we can identify a sort of passage from a still fully modern conception of the elements involved (i.e. "the human", "nature" and "things"), in the transition from phenomenology to later accounts. I am not going to directly address all the main issues related to the "essence" of things *an sich* (quoting a Kantian expression), for it would imply a long digression that would lead us way too far from the strict topics of this essay. Thus, for a deeper understanding I shall recommend referring to Merleau-Ponty's critique of Sartre and of the limits of phenomenology (Merleau-Ponty 1962; Ronchi 2014; Van der Veken 2000).

But just to make things clear, these limits are precisely referred to that *an sich* of things that phenomenologists like Sartre ended up reconfirming while trying to get rid of. Thus, quoting Dolphijn and Van der Tuin (2011), with regard to phenomenology we could say that "any negation results in an unwanted reconfirmation of the negated", i.e. the alleged essence of reality (Hoel and Van der Tuin 2013, p. 191). On the other hand, Giovanni Leghissa, among others, criticized Heidegger for not being capable of interpreting science and technicity as being part of that "natural history". Referring, for example, to Heidegger's *Das Ding* essay, Leghissa explains how the philosopher, still from within a fundamentally humanist framework, is "unable to think of the history of being without a reference, at least implicit, to the *Geist*" (Leghissa 2014, p. 15, translation of the author). For this reason, "Heidegger places the manifestation of the *Geviert* in the pitcher, magically enclosed in a simple daily gesture such as pouring water or wine", reinforcing "the idea that experiencing the world through the mute trade with objects leads mortals to approach the sphere of the divine. Consequently, [...] manipulating the jug precisely marks the abysmal distance between humans and animals" (idem, p. 16). Nevertheless, in his early work (*Sein und Zeit*) Heidegger actually investigated how the world is given to humans through the *praxis* of their everyday relationship with objects. Thus, we can agree with Leghissa and Van der Veken when saying that the German philosopher went close to taking that fundamental step of reading technicity as a feature of natural history, but apparently got lost in the poetry of the rituals of object-related gestures to address the problem from a different perspective before he died. Thus, it is in the work of other authors that we can identify a more radical openness towards the role of technicity in defining "the human". To this extent, I identified in some postphenomenological and "continental" (referring to the label of continental philosophy) approaches potential alternatives for interpreting our relation with technology.

Talking about the inclusion of technicity we see how, even if not explicitly, the concept of heteroreferentiality emerges, and this is what I would like to emphasize. In other words, "the human" taken for itself, rather than being a measure of all things, is just an incomplete element of a bigger equation, the ongoing process of co-formation that requires "nature" and "the artefact" to express its potential. Thus, the point I would like to make is that we cannot get rid of our belonging to nature, because our alleged superiority in comparison with other species is a wrong interpretation precisely of our genetic, i.e. natural, potential. Consequently, technical skills are not tools to be exploited, but the expression of such potential, as much as the basket making skills are the expression of the peculiar potential of a weaverbird (Ingold 1997, p. 113). Ever since Darwin demonstrated the existence of a red thread connecting all living beings, all the steps forward that we have taken in the direction of a better understanding of life have been confirming this, from both an archaeological and a genetic perspective. Recalling Tim Ingold's reflections upon technology, through the example of the comparison with the weaverbird's skills, "in both cases we are dealing with a skill that is neither innate nor acquired, but developmentally incorporated into the modus operandi of the body—whether avian or human—through practice and experience in an environment. There seems, then, to be no clear-cut distinction, after all, between the skilled making of the animal and that of the human" (ibid.). To my understanding, the recognition of the role of technicity in our existence as individuals and as species passes through the acceptance of our—peculiar—belonging to nature.

3.2 Culture Is Nature

Aud Sissel Hoel and Iris van der Tuin exposed how a possible alternative path has been developed by thinkers such as Ernst Cassirer and Gilbert Simondon. Both the philosophers "take their point of departure in a shared perception of a failure on the side of thinking, both scientific and everyday cultural, to grasp the true nature of technology. Antagonisms in culture flow from this ignorance, which manifests itself in technophobic reactions that configure the machine as a 'foreigner' devoid of human reality or, alternatively, in untempered technophilic reactions that exploit its technocratic potential, or that seek to enhance the human being or replace her with better-functioning doubles" (Hoel and Van der Tuin 2013, p. 190). Within the last lines we can see how the critique is directed explicitly to those forms of contemporary "reincarnations" of humanist beliefs, in their only apparently contradictory opposition. Ignoring the true nature of technology, as suggested from the beginning of this essay, can thus take two apparently opposing derivatives: technophobic or technophilic.

For the two philosophers, nature "is not a fixed category. The human has no direct access to the form of nature, nor does she impose this form by her mind" (idem,

p. 188). Similarly, the notion of "human" is not rigid. "Since nature is not a stable agent, it does not keep a firm hold on the human. Instead, the human is seen as deeply entangled with nature, engaged in ongoing processes of co-formation" (ibid.). This last sentence closely resembles the thoughts of the aforementioned French philosopher Merleau-Ponty: "meanings are truly 'con'-stituted in the interplay between ourselves and the world" (Van der Veken 2000, p. 323). With regard to this interplay between the human and the world, Cassirer and Simondon perceived the importance of adding "an irreducible third ingredient in the ontological entanglement: technicity" (Hoel and Van der Tuin 2013, p. 188). Acknowledging the constantly ongoing process of co-formation intervening in human-nature relationship, "the human no longer possesses the exclusive power to define" (ibid.), and this paves the way for the entrance of technology into the "equation". If we were to follow Henri Bergson's position, not only technology would find its place within the constitution of "the human", but it would stand as the leading feature: "Intelligence, considered in what seems to be its original feature, is the faculty of manufacturing artificial objects, especially tools to make tools, and of indefinitely varying the manufacture" (Bergson 1998 [1911], p. 139). Technological objects put both the notion of human and that of culture in motion acting as a creative mediator in terms of their functioning and ontological force. The latter term represents the fact that the renewed role played by technological beings is no more that of passive objects in the mere material meaning of the term. We can speak about ontological force precisely because those beings contribute to shape, in an ongoing and mutual process, the ontological category of "the human".

Interestingly, Simondon uses the expression "genesis" when referring to technical objects (while this is typically the exclusive prerogative of living beings), to get rid of the *hic et nunc* thingliness typical of other objects, turning technical ones into "units of becoming" to express the process of "differentiation and refinement". On his side, also Cassirer insists that we cannot pretend to grasp the essence of technology from a perspective that interprets its being as "mere things with properties" (Cassirer 2012, p. 32). It is through leveraging on the processual aspects, i.e. activity or function, that we can successfully understand their philosophical relevance. It is the processuality, made up of recurrent causalities, that forces to include not only objects themselves, but also makers, users and the environment in the "con"-stitution process. Recalling Bergson, what we intend when talking about "the human" and his peculiar form of intelligence arise from this constantly ongoing creative dialogue, rather than being a property of the Cartesian *res cogitans*. The example of the "substrate-independent mind" (SIM), in the previous section, was precisely intended to demonstrate how current transhumanist position has only apparently moved forward in comparison with classic humanism.

3.3 The Constitutive Role of the Technical Mediation

To keep the elements of the equation in motion, and to avoid any possible essentialist interpretation, another fundamental concept is that of technical mediation within the process, i.e. the expression of the ontological force that things exercise. Cassirer gives "positive account of mediation as, at once, invention and intervention [. . .]. In the resulting account, the in(ter)vention of a 'foreign' symbolic or technological apparatus is seen as a prerequisite for the discovery and disclosure of nature, and not as an obstacle to it" (Hoel and Van der Tuin 2013, p. 195). Intervening "into the free rhythm of natural movements" (Cassirer 2012, p. 40), the status of the technological mediator moves from the secondary role of substantivist metaphysics to a constitutive one. Cassirer interprets the intervention of the mediator as it temporarily disrupts the natural flow of movements. From his perspective, this allows "these movements to resettle on a new and 'higher' or more articulated level. The tool's articulatory intervention is also an *invention*, since it occasions a metamorphosis into something new. This in(ter)vention should be understood as making its mark *with* and not at a distance from or as a distancing from natural movements. It is in this sense that there is nothing humbly 'natural' about the movements nor is there something authoritatively 'cultural' about the work with the tool" (Hoel and Van der Tuin 2013, p. 195). As the authors pinpoint, the action of the technological mediator does not really break into "the free rhythm of natural movements" from the outside because, for Cassirer, interpreting the invention/intervention process as a distance from the natural movements paves the way for alienation. Referring to this last point, Simondon raised a relevant observation that may help to grasp what his conception of the nature of technical objects has to do with the notions of automatism and of machine development. We already saw that technical evolution should be considered an ongoing process, but it could be less intuitive to state that such process does not lead to an increased automatism (which is a strong cultural belief). Simondon motivates this assumption saying that a completely automatic machine does not tend to perfection. Instead, it is limited in its functional possibilities by its predetermined functioning and by being closed upon itself. Simondon affirms that "a certain margin of indetermination" (Simondon 1980, p. 4), and therefore of openness, is required for a machine to evolve to more sophisticated levels. The crucial point is that this openness is what allows the machine to receive and assimilate information, which means being capable of connecting with other systems, including the human and, eventually, nature. For a technical object to be considered as "evolved", the human cannot be "outside" the machine itself; he must be, instead, "among the machines that work with him" (ibid.). And again, concerning nature, such machine incorporates "part of the natural world which intervenes as a condition of its functioning and, thus, becomes part of the system of causes and effects" (idem, p. 46). Becoming an active agent in the world creation process, the technical object does not just entertain a relationship with the existing environment, it cooperates in creating a new system. This leads the philosopher to affirm that the system precedes form and matter which "are at the same level, are part of the system; there is continuity between the technical and the natural" (idem, p. 244). The constant process of redefinition of the environment or in(ter)vention, which on the side of the technical object Simondon calls concretization, involves modifications also on the sides of the human (as inventor and end user) and of nature. As it was for Cassirer's notion of foreignness, the mediator rises to the role of ontological force and challenges the substantivist notions of "nature" and "the human" to put them in motion.

The relevance of this idea of technical mediation emerges even more explicitly in recent accounts such as postphenomenology and the so-called material entanglement theory (Ihde and Malafouris 2018), where its evolution is pushed to the extreme: "humans and things exist in mutual interdependency, beyond the nature and culture distinction" (idem, p. 5). Thus, our peculiarity as species, which for the sake of comprehensibility is what we might call "technical intelligence", is a product of evolution by means of selective pressure, but in a way that would not be possible without us mingling with things from the very beginning. Thus, the categories are put into motion, but a creative, enacting motion. Here, we can detect another meaningful difference from even the most moderate transhumanist accounts: the mediation is constitutive and embodied, not a mere and superficial interaction. Co-evolution is not adaptation. Thus, the enthusiasm towards technological enhancement typical of neo-humanist currents appears to be totally misleading in explaining the kind of engagement we entertain with the materiality of the world of things. Following Ihde and Malafouris rephrasing the concept of *Homo faber*, it "*does not refer to a special ability that only humans have*, rather it refers to *the special place that this ability has in the evolution and development of our species*. The difference that makes the difference is not just the fact that we make things. The difference that makes the difference is the recursive effect that the things that we make and our skills of making seem to have on human becoming" (italic in the original, idem, p. 4). Reconnecting this point to Merleau-Ponty's intuition, we can say that human beings and things are "inseparably intertwined and co-constituted" (ibid.), and this was valid in prehistorical times as much as with the emergence of digital and computational cultures. Besides, despite what humanists may believe, "there is no 'core' or 'essential' humanity (biological or other) that pre-exists and which could subsequently be enhanced, extended, disciplined or threatened by technological interventions" (idem, pp. 4–5). Even if we were to put it in those "technocratic cyborgian" terms that are so important for transhumanists, paraphrasing Latour's popular book title, *We Have Never Been Modern*, using Bernard Stiegler's words, we can say that we have always been cyborgs (Stiegler 1998). Technology has always been constitutive for our species, on a level that goes far beyond mere physical interaction.

Thus, from within this renewed interpretative framework the focus is to be put on the recursive movements in which nature, the human and technology are all involved in the constant world creation and discovery process. There is no more need for fixed and solid relations coming from the outset, but a recursive and reflexive, perpetually in(ter)ventional movement of concretization/formation through differing (using respectively Simondon's and Cassirer's terminology) that is aimed to enlighten the ontological force of technicity, and to become a real alternative to both substantialist and subject-centred accounts. We can leave this part of the discussion going back for one moment to one of the main questions addressed at the beginning of this essay, saying with Simondon that, if we do not want to hide behind a blinding "mask of facile humanism", a theoretical approach (based upon the cultural background of the Western world) aimed at understanding the impact of technological development on our society, we must decidedly not just take into consideration, but also assimilate, technological beings. This way we can do well in understanding the relationship between the human, nature and technicity without having to refer to any apparently divergent technophilic and technophobic approaches.

4 Conclusions

From the starting assumption of the inadequacy of the modern humanist background in terms of analysing the impact that technology is having on our society, I looked for potentially more inclusive alternatives still within the Western tradition. Trying to reshape the leading mindset by showing its intrinsic limits represents a gigantic effort, but after philosophy paved the way, other disciplines followed, contributing with their peculiar approach to create a different framework within which the epochal change driven by an apparently unstoppable technological growth is interpreted. Besides the mentioned contribution derived from neuroscience, we can see how many other fields of research (from sociology, to archaeology and anthropology, just to give an example) are, at least partially, shifting towards a more inclusive perspective, in regard to technological otherness. Apparently, following such a non-linear path can provide us with a more reliable interpretative framework, which is not informed by the traditional ontological categories and thus can give birth to a more inclusive epistemology. Thus, my hope for the future is that, rather than letting technophobic and enthusiast approaches rule the debate, such a multidisciplinary approach will become the standard, especially when it comes to a delicate issue such as the impact that technology is having on our society.

References

Bergson, H.: Creative Evolution. Mitchell, A. (Trans.). Dover, New York (1998) [1911]
Carli, E. (ed.): Cervelli che parlano. Il dibattito su mente, coscienza e intelligenza. Paravia Bruno Mondadori Editori, Milano (2000)
Cassirer, E.: Form and technology. In: Hoel, A.S., Folkvord, I. (eds.) Ernst Cassirer on Form and Technology: Contemporary Readings, pp. 15–53. Palgrave Macmillan, Basingstoke (2012)
Damasio, A.: Descartes' Error. Emotion, Reason and the Human Brain. Penguin Books, New York (2005)
Dennett, D.: Darwin's Dangerous Idea. Evolution and the Meaning of Life. Simon & Schuster, New York (1995)

Dolphijn, R., van der Tuin, I.: Pushing dualism to an extreme: on the philosophical impetus of a New materialism. Cont. Philos. Rev. **44**(4), 383–400 (2011). https://doi.org/10.1177/1077800416673660

Edelman, G.: La mente, una prospettiva evoluzionista. In: Carli, E. (ed.) Cervelli che parlano. Il dibattito su mente, coscienza e intelligenza. Paravia Bruno Mondadori Editori, Milano (2000)

Galimberti, U.: Psiche e techne. l'uomo nell'età della tecnica. Feltrinelli, Milano (1999)

Hoel, A.S., Van der Tuin, I.: The ontological force of technicity: reading Cassirer and Simondon diffractively. Philos. Technol. **26**, 187–202 (2013). https://doi.org/10.1007/s13347-012-0092-5

Ihde, D., Malafouris, L.: *Homo faber* revisited: postphenomenology and material engagement theory. Philos. Technol. **2018**, 1–20 (2018). https://doi.org/10.1007/s13347-018-0321-7

Ingold, T.: Eight themes in the anthropology of technology. Soc. Anal. **41**(1), 106–138 (1997)

Ito, J.: Why westerners fear the robots and the Japanese do not. https://www.wired.com/story/ideas-joi-ito-robot-overlords/ (2018)

Kiberd, R.: Burn it all down: a guide to neo-luddism. https://gizmodo.com/the-many-faces-ofneo-luddism-1682139778 (2015)

Kurzweil, R.: Promise and peril. http://www.kurzweilai.net/promise-and-peril (2001)

Leghissa, G.: Ospiti di un mondo di cose. Per un rapporto postumano con la materialità. Aut. Aut. **361**, 10–33 (2014)

Marchesini, R.: Post-human. Verso nuovi modelli di esistenza. Bollati Boringhieri, Torino (2009)

Merkle, R.: Uploading. In: More, M., Vita-More, N. (eds.) The Transhumanist Reader. Classical and Contemporary Essays on the Science, Technology, and Philosophy of the Human Future, pp. 157–164. Wiley, Chichester (2013)

Merleau-Ponty, M.: Phenomenology of Perception. Smith, C. (Trans.). Routledge & Kegan Paul, London (1962)

Moravec, H.: Mind Children: The Future of Robot and Human Intelligence. Harvard University Press, Cambridge, MA (1988)

More, M.: The philosophy of transhumanism. In: More, M., Vita-More, N. (eds.) The Transhumanist Reader. Classical and Contemporary Essays on the Science, Technology, and Philosophy of the Human Future, pp. 3–17. Wiley, Chichester (2013)

Parisi, D.: Mente. I nuovi modelli della vita artificiale. Il Mulino, Bologna (1999)

Ronchi, R.: Figure del postumano. Gli zombie, l'onkos e il rovescio del Dasein. Aut. Aut. **361**, 82–96 (2014)

Simondon, G.: On the Mode of Existence of Technical Objects. Mellamphy, N. (Trans.). University of Western Ontario, http://dephasage.ocular-witness.com/pdf/SimondonGilbert.OnTheModeOfExistence.pdf (1980)

Stiegler, B.: Technics and Time 1: The Fault of Epimetheus. Stanford University Press, Stanford (1998)

Van der Veken, J.: Merleau-Ponty and Whitehead on the concept of nature. Interchange. **31**(2&3), 319–334 (2000). https://doi.org/10.1023/A:1026764822238

Whitehead, A.: Process and Reality, Corrected edn. Free Press, New York (1978)

Chapter 4
Senseless Transhumanism

Tomáš Sigmund

Abstract This chapter analyses the hidden presuppositions of the transhumanist movement and asks if transhumanism can contribute to the revelation and broadening of the sense of the world. It uses the Arendtian perspective of three human activities (labour, work, action) and places transhumanism into their context. In order to do that the concept of intentionality is applied to show the difference between man and machine and the deficiencies of machines in terms of sense constitution.

1 Introduction

Earth and cosmos are man's home. However, since the beginning of modern age man is not content with his world anymore and tries to flee from it. World has been reduced to the condition and necessary evil of man's existence which must be transformed through labour to provide resources for human life. The never-ending striving for progress, effectiveness, accumulation, growth, etc. is the example of an infinite automated process deprived of human understanding and sense. The whole society has been subordinated to the laws of nature or laws of history. Men have become slaves of their laws; they know labour and its opposite—leisure in the form of idleness—only. The human condition is considered loaded with necessities and duties. Transhumanism as a movement to enhance human physiology and intellect (Bostrom 2005) aims at providing man with the full autonomy, removing every necessity and barrier to his free will. However, a movement continues in this tradition of the senseless world.

T. Sigmund (✉)
Prague University of Economics and Business, Prague, Czech Republic

W. Hofkirchner, H.-J. Kreowski (eds.), *Transhumanism: The Proper Guide to a Posthuman Condition or a Dangerous Idea?*, Cognitive Technologies,
https://doi.org/10.1007/978-3-030-56546-6_4

2 Three Man's Activities

According to Arendt (1958), man is endowed with three types of activities. The first one is labour which allows man to meet necessities for self-preservation and reproduction. These needs require continuous satisfaction and so labour is a never-ending process with vanishing almost unstable results. Because labour is directed by necessity, the labourer is similar to a slave who has no freedom. The whole realm of economy belongs into this area. The counterpart of labour is consumption. Because necessity leaves no space for freedom and creativity, labour is done in isolation as well as the consumption of its products.

The second activity is work, which leaves a durable object that is not a direct object of consumption. These objects create the world we live in. Work is guided by the instrumental reasoning which looks just for instruments for its ends. Work disturbs the natural character of man as it is an act of violence on nature and so proves the non-animal character of man. The prevalence of use value over worth in economic thinking is a symptom of this activity. Work exhibits a kind of freedom and is public because it creates a common world, which stands between humans and unites them at the same time. Its products are preconditions for the existence of a political community. Unfortunately, in the economic society the characteristics of work products: permanence, stability and durability were exchanged for life subsistence, productivity and abundance.

Work is still subordinated to its instrumental character. Work is a necessary instrument or means for the production of a product. Work cannot be free as it is not an end in itself; it is subordinated to other ends. However, man is also capable of a third activity where he can express his full freedom and which is not subordinated to anything else: action. Arendt considers inner freedom to be only derived from practical freedom experienced in the intercourse with others. Men are free as long as they act. Action means the ability to begin something new. Action requires others to see it and to give it meaning. Action needs a public space where it is realized and which allows individuals to encounter one another. Action includes both speech and action and allows man to disclose himself to others and to distinguish from others as a unique individual. Action is directed to other human beings and generates relationship. It is watched, interpreted and evaluated by others which gives action its durability and non-instrumentality.

Due to historical development framed by three historical events (discovery of America and exploration of the whole earth; reformation and related expropriation and accumulation of social wealth; the invention of new scientific instruments and development of science that considers earth from the perspective of the universe), our alienation from the world increased according to Arendt. The disappearance of distances on Earth through exploration and transportation has interrupted our relation with one place on the earth, expropriation has interrupted our relation with the land and its place on earth, and the perspective of modern science has treated earth from the perspective of the universe which has interrupted the dominance of the earth-centred view which is natural for us.

Arendt's differentiation of human actions may serve just as an instrument for understanding human activity, none of the three activities are performed isolated from the rest as Arendt puts it, but they may be considered as representing some tendencies existing in human behaviour. Their analysis shows how we have neglected work and action and supported labour instead. The result is the loss of our place in the world and loss of meaning and sense that we find in the world. Transhumanism attempts to liberate man from this miserable situation by removing all the necessities from human life. However, the sense of the world is not recovered, because there would be nothing to limit man in his arbitrary behaviour guided by nothing meaningful, only by consumption. Transhumanism therefore extends the area of labour or possibly work, without effects on the sphere of action.

3 Intentionality

The reason why machines cannot help us give sense to our world is that they do not have intentionality. J. Searle (1984b) points out the specificity of human mind which is not tractable by natural sciences focusing on the physical world. Mind is conscious, free, mindful and rational in contrast to mindless physical particles. Human mind has four important features of consciousness, intentionality, subjectivity (my perceptions are not accessible to anyone else) and mental causation (our thoughts influence the way we behave and thus have a causal effect on the physical world). The mental causation seems to be an addition, but actually it is a higher degree of intentionality and is related to free will. Free will means that our behaviour is not predictable the same way physical phenomena are predictable. Men have an experience of free will. One can always falsify the prediction somebody makes of him. Free will is always related to consciousness; only conscious beings can have free will. Free will means the experience of engaging in voluntary intentional human actions (Searle 1984b, p. 91). We must be aware of alternative courses of actions in order to experience free will. The experience that I am doing something contains the experience that I could be doing something else. Acting includes the experience of freedom. We cannot erase freedom from our vocabulary as it is used to identify and explain behaviour. It must be differentiated from arbitrariness; the deciding person must know and agree to the arguments and is responsible for the decision.

Searle (1984b) concludes that artificial intelligence lacks semantics and that is why no computer program is by itself sufficient to have a mind. Minds have mental contents, more precisely semantic contents, whereas computer programs are defined solely by formal or syntactical structure. I think all four features of the mind are based on the concept of intentionality.

In the article from 1984 "Intentionality and its place in nature", Searle (1984a) gives more details on the concept of intentionality. He analyses four sentences two of which are relevant for our discussion of intentionality: "Bill sees that it is snowing", "My car thermostat perceives changes in the engine temperature". In the first sentence the intentionality is correctly ascribed. In the second one, we can speak of intentionality metaphorically only as thermometers do not have any perceptions. The intentionality has been only transferred from an agent who uses thermometer to regulate the temperature of the engine. The thermometer can only react or respond to the temperature measured. And that is the way machines work.

Contemporary discussions on the concept of intentionality are related to Franz Brentano who defined intentionality in relation to scholastics as "reference to a content, direction toward an object, . . ., or immanent objectivity" (1874, pp. 88–89). Mental states include in his understanding an object within itself. His pupil and follower E. Husserl made this concept the principal theme of phenomenology. Consciousness has the feature that it is not just affected by things and does not just react; it is also conscious of things. Many mental states have the feature that they are about something. McIntyre and Smith (1989) call it the representational character of consciousness as each mental state is a representation of something other than itself. The reduction of intentionality to a causal relation is not acceptable as experiences like imagination, hope or experiences of imaginary objects have no causally effecting part. The object of an intentional state is not necessarily an external object. Intentionality is therefore a property that mental states have independently of the causality of the external world. And what is more, intentionality changes internally, independently of the object. The relation to an object changes as I know more and more about it, but the external object remains the same. Intentionality is therefore independent of the existence of its object and dependent on the conception of the object (McIntyre and Smith 1989). We know intentionality predominantly from the first-person perspective and from the subjective perspective, and it is impossible to explain it from the third-person perspective, claims Husserl.

The content of an act makes our act a representation of an object which we are aware of. We are not usually aware of the content, of the sensual data, but of the object—the thing we perceive, desire, fear, etc. As a result, intentionality depends on the content of the act which is independent of an external object. In addition to that, different contents can be directed towards the same object and represent it differently. And now we are coming to the crucial point of Husserl's intentionality which is relevant for the discussions of transhumanism. To simplify matters a little, for Husserl sense or meaning is that which provides an act with its intentional character. An example can be language where meaning is given to various sounds to speak about extra-linguistic things. So language is similar to intentionality as it is also directed towards something. Different words can represent the same referent in different ways. And the linguistic reference of words is not dependent on their objects, but on their meaning. The sense of the act gives the subject a sense of the object and puts it in relation to it. The sense prescribes the object.

The way we are intentionally related to the world and represent it is influenced by our bodily existence as Merlau Ponty has explained. The movements of the body put things into their context and include them into its perspective. We understand things as being on the right, on the left, etc. in relation to our body.

It follows that we still cannot build machines that would surpass our intentional abilities as they work on the basis of syntactical reactions and are not capable of intentionality and its semantics. Intentionality can have different abundant features and levels. We can build labouring machines only that would liberate us from the necessary actions; at most they could be used as instruments fulfilling our aims, but they will have no aims of their own in the strict sense of the word. They would not be capable of creativity and would not be able to give sense to our lives as they would not be able to interpret the world and so give it meaning.

4 Transhumanism and Utopias

Transhumanists concentrate on the removal of any necessities in man's life. This tendency is one of many perspectives on human nature; it is not the only one. Julian Huxley (1968, p. 17) defined transhumanism as an effort to overcome the limitations of human nature. Transhumanists even want to control evolution. These tendencies are quite frequent in utopias. Some of them show how one-sided transhumanist ideology is and what type of perspective on the world based on intentionality we would lose. Utopias are pictures of a perfect world where perfect human beings live in a perfect social, political and technical environment. The idea that we will be able to fulfil all desires and live life full of joy is known from mythology. Transhumanism focuses on sensual pleasures, youth and no restrictions of freedom. People will live in the Golden Age described by Hesiod (Hauskeller 2016, p. 15). The emphasis on enhancement of cognitive abilities is similar to alchemists. The tradition of techno-utopian thinking goes back to the ideas of F. Bacon, and in the twentieth century, this tradition was followed by H.G. Wells who stressed the constant progress in modern utopias. In that he followed the tradition of the French Revolution (M. de Condorcet) who believed in indefinitely perfectible humans within the boundaries of human nature (Hauskeller 2016, p. 18). The transhumanist arguments are therefore based on old utopian ideas that have long histories and are rooted in our imagination. There are not only rational arguments in transhumanist depictions, but values, perspectives and ideas which shape what is the goal of human life, who is man, etc.

The overcoming of limitations of human nature should lead to complete autonomy. The risk may be that bodies will become objects of fashion rather than grounds of being as Hayles (1999, p. 1) puts it. The idea of overcoming all limits and boundaries is utopian. Transhumanist utopias are framed by dystopian descriptions of the current state of affairs which suffers from death, illnesses, ageing, etc. Utopian descriptions are a combination of promises and commands. The promises of a better future are conditioned by certain ways of collaborative behaviour. Transhumanists claim it is our moral duty to improve the human nature. Hauskeller (2016) concludes that the old ambitions to transform human nature gave rise to the technical solutions allowing it. Science is driven by non-scientific purposes. That is not bad; dreams are very useful in driving and directing human efforts, but we must not forget that not all dreams are worth dreaming or realizing. Every radical transformation is dangerous and those governed by utopian thinking are risky as they promise something they cannot fulfil and in fulfilling their promises they cause many harmful effects. We do not know whether the promised transformations will work or not. We should not forget that every paradise has a dark side, that there is always a price to be paid for everything and what seems to be a progress may in fact be a downfall (Hauskeller 2016, p. 202). In other words, transhumanists let things appear in a better light than what is truth, and this they have in common with utopian descriptions. The benefits are depicted so gloriously that nobody can resist.

The second problem with transhumanist thinking is that it shows the progress towards new human conditions as inevitable and included in the human nature from the beginning of mankind. We can slow the progress down, but we cannot stop the movement, transhumanists claim. The positives of the future depicted in transhumanist stories make our present situation seem almost unbearable.

The question of what it means to be a human is related to the ideal of the human nature which often has a utopian character. The utopias of an ideal human society show qualities that we do not, but should have and so serve as a mirror in which we are ashamed to look. The transhuman beings can claim to be the real humans, which comply with the human ideal. Every attempt to define human nature is an attempt to define what men should do and how they should behave. To define the human nature is to tell a story: about what it means to be a human, about what is worth doing, desiring, fighting for, about good and evil and what life is all about (Hauskeller 2016, p. 49). J. M. Smith (1998, p. 375) claims that an important function of myths is to "give moral and evaluative guidance" and to provide "a source and justification for values". People tell myths in order to "persuade others to behave in certain ways". Myths therefore tell us how to behave and what to do. Human nature may be indefinable, but we need guidance to make sense of our lives. There are many differences between humans, but also many similarities. It is not clear what we should concentrate on and no definition can grasp man in his totality. But still we need an ideal to give structure and light into our world (Hauskeller 2016, p. 52). The problem with transhumanist ideal is that we have used one ideal of the animal laborans only and forgot other dimensions of human existence.

5 Loss of Sense in Transhumanism

Transhumanists place a lot of attention to escaping from the physical which is considered a burden. It resembles a religion which tries to escape the bad world. However, transhumanists prefer action over thinking, develop engineering solutions and only afterwards think about their implications.

Removing some of man's limits like death, illness, tiredness, etc. and creation of other limits may change understanding of our life and its fulfilment. We strive for something, work, learn, fear and without the appropriate context man will decline as he would lose sense of his world (Kass 2003). The limits are not only barriers to our freedom, they can also be interpreted, we can give them meaning, they can determine our lives. Because this dimension is missing in transhumanist thinking, it concentrates on the removal of any barriers and necessities as they are considered a burden in the physiological sense.

If humans had better abilities and more advanced bodies than we have, it is not clear whether they would also have higher moral status and whether they would be justified to sacrifice their lives and living conditions. There is a certain hierarchy in our thinking. Physical objects do not feel and so have only very little moral status. Animals have some moral status because they can feel if somebody hurts them. However, they do not have higher cognitive abilities and so we can kill them without special justification. The existence matters to humans, they can feel and think and so we cannot kill them. Humans have autonomous existence, are free persons and so have a different kind of dignity and value to other beings. However, if for some reason which need not consist in their higher cognitive abilities the transhuman beings believe they can use us for their benefit, we could not do anything about it. They need not recognize our moral values and we would not have instruments to persuade them to do so.

According to transhumanism our aim is to become better than we are now. We have not achieved our aim yet. We live in deficient conditions, we do not have enough control over the world including our bodies, we experience too much pain, our lives are too short and end in death which should be avoided. In order to make world worth living, we should become different (Hauskeller 2016, p. 56). Our situation is in transhumanist view new, because it is nothing external that limits our nature, it is the limit of our abilities, we are facing limits of our nature, our inabilities. Nothing external is limiting us; it is us who have the limits in ourselves. Our nature is not our ground, but our barrier which we perceive in the form of illness, death, age, emotions, lack of intelligence, etc. Human nature must be improved and removed if possible because it limits us. Human nature is in transhumanist view very much identified with our body and that is why its elimination is related to the removal of corporality. Ideal human being is a being without body. However, if we interconnect our bodies with a machine and grant it better abilities, this new body (1) will not be ours, but it will be something external, and (2) we would always be dependent on its proper functioning. There must be somebody providing maintenance and we will never have such instruments under full control. Our relation to machines is different to our relation to our bodies and because we have a closer relationship to our bodies we feel ashamed when we compare our bodies to machines. Our bodies are vulnerable, imperfect, can be destroyed, etc. They are also not very nice inside, in its inner content and structure. They are very fragile and we are afraid of them and their limits. We are, however, afraid of machines, too, as they are an unknown territory for us.

In the past, ancient or medieval automata were designed to imitate features of living beings (Hauskeller 2016, p. 63). So the first intention was to imitate the perfect nature. Machines adapted to human beings. However, nowadays we adapt to perfect machines. Hauskeller (2016) distinguishes four stages of man's convergence to machine. (1) Illusionism where the appearance is changed and improved, like in cosmetic surgery. But the problem is not solved; it just must be periodically repeated. Time is not stopped, only skipped. (2) Fortification of physical enhancement aims at strengthening the body, making it less vulnerable and more capable, like the exoskeleton which gives us more power and speed. We retain our body and add another more powerful to it controlled by the physical one. (3) Replacement of parts of our body with artificial ones that have better abilities and characteristics. Every part of our body can be replaced. (4) Displacement tries to remove the physical body completely by uploading the mind to a computer. This perspective looks at our body as something deficient and burdening. However, we may look at our body also positively as something that allows us to do some things. It is not capable of everything, but that is not possible; any ability is a limitation at the same time. Omnipotence is a fantasy. We can enjoy the things we can do instead of dreaming of things we cannot do. Hauskeller (2016, p. 71) says that death is a misery, but also allows life to continue in new ways and fresh eyes. To turn bodies into machines is not so urgent. A healthy body is a big gift, which must be taken care of, but not necessarily replaced with a machine.

To understand nature as our limit in the form of our body is not the only way how to understand it. This transhumanist use has negative connotations, but it is not the only way humans can be defined. There must be something more in humans that forces them to proceed further. And that is the second meaning of nature, as something that enables us to trespass our limits, to go beyond them. It is our nature to turn against nature that limits us. In Nietzsche's word "man is as yet undetermined animal" (1966, WII, p. 623). We can be in such ways that no other animal can and we should pursue it. We are not yet determined and we can decide who we shall be. Our nature is to have no nature as Mirandola puts it (1985, p. 4). Being content with what we have would be a betrayal on our nature. We should form the world and form ourselves. We should not contemplate; we should improve our lives. And the aim seems to be not any particular aim including the moral nature, but freedom from any determination, possibility to do and be anything. K. Jaspers (1971) says ideals of man are always deficient because they lack universal validity. That is also what the political realm teaches us.

Transhumanists use these ideas not for the development of creative or political powers, but as arguments for prolonging human life into eternity; the point is to free oneself from the limits of ageing and from the limits of being one particular person. Unending life would allow us to start over and over. Their freedom is, however, understood as an absence of limits, i.e. negatively only, whereas both creative and political freedom mean freedom from something and/or to something. Freedom as an absence of limits is empty.

In any case it is illusionary to think that we could get rid of our past life completely. The belief behind this attempt is that with technological improvement men will be better people and happier and lead more fulfilling lives. How short-sighted this belief can be has been proven by many tyrannies and totalities. Many transhumanists stress the continuity of Enlightenment's values which must be fulfilled in the future. Transhumanists think that our main characteristic consists in our mind, in our thinking that should be uploaded to a machine one day. Saving lives in any form is transhumanist priority which means eternal life is implicitly positive. We may, however, ask if the life of everybody deserves to be prolonged. Having eternal life extends the position of man in the middle of the world which is revolving around him. We must learn to die as well as we must learn to live our lives. A lot of people live too long, says Nietzsche (1966, WII, p. 333). Immortality means in his perspective denial of differences and ranks between men and pretends we are all equal. It also substantiates individual selves. It allows one to become more important than the rest of the world and to care just for himself. The self just takes care of itself and the world has no importance for it. The most important thing is the survival and well-being of the self. This is the indication of impotence for Nietzsche. That is why "It needs to be overcome" (1966, WII, p. 303). Dying means entering the ocean of becoming. The overhuman knows how to live and how to die. The transhumanist does not know how to live and so neither how to die.

If we accept two types of human nature, one identified with our body and limiting and the other identified with our mind and will and breaking any limits, we will end up in a dualist conception of man. If we refuse the dualistic thinking and accept limitations as something that gives us identity and values and opens possibilities, we will come out of the transhumanist thinking. We should accept that we can do a lot of things because there are many things that we cannot do. There are many things which are not under our control and that is why we value them (love, health, happiness, friendship, experiences, etc.). The problem with transhumanists is that they see just one part of the picture which suits them and is in compliance with their perspective. There should be enough freedom and space for those who decide not to enhance their bodies in the way transhumanists want them to (Hauskeller 2016).

We also should not forget that the power we gain through new technologies will be their power which can be lost and so we will be more dependent on them. It will not be our power; it will be just borrowed. Every increase in power increases vulnerability at the same time, be it vulnerability of some part of mankind or vulnerability of all the mankind in relation to technology. And the values that once stood at the beginning of new technologies would become controllable at the end as the designers would be able to decide what kind of conscience they will produce (Lewis 1955, p. 74). On what basis will they decide? Science cannot answer questions on its own. So every decision will become arbitrary, without reason. As a result, man will not conquer nature; he will not act freely and according to reasons, but according to his nature. And so nature will control man as Lewis points out (1955). If we make ourselves controllable, we will treat ourselves as raw material, as something controllable by the designers and producers and their nature. A good example can in this case be various manipulative techniques that give oneself power over others, but also increase his vulnerability. The more power we have over the world, the more power the world has over us. And only the designers can on the basis of their arbitrary decisions decide what direction the development trajectory will have, what will be developed and how.

M. More (2013) emphasizes progress and activity in creating better future rather than faith in it. Transhumanism believes in the increase of individual autonomy and whatever decreases it should be avoided and fought. Technologies enhancing humans support autonomy and that is why they should be supported. This is the key point in the transhumanist mythology (Hauskeller 2016, p. 98). If one does not think that individual autonomy is the most important thing and that people should not work on their self-creation, one is not a transhumanist. So autonomy is something that should be achieved and in this sense it is a value, not a fact. All limitations are wrong including the limitations of the person and self. It may well happen that we would become part of a universal self and lose our identity.

Our limits are as a matter of fact related to the nature in us. Breaking the limits means getting rid of the nature in us, killing the animal in us. The sick ones, even animals can be helped, too, and promoted to the rank we have and then even further where we strive ourselves. What is shown here is pity and sentiment, which is an expression of superiority. We consider ourselves better than the poor and consider even our position poor and miserable from the transhumanist perspective. Human is a better animal and transhuman is an even better animal. But human or animal can be better in some respect and for some purpose, not per se. There is no paradigm for the human or animal; there are just purposes which can be served better or worse. Both humans and animals are too complex to follow just our orders; they cannot be simply regulated and controlled. They are more than just regulated reactive machines following certain roles. Hauskeller differentiates two kinds of freedom (2016, p. 116). One is the self-governing autonomy consisting of liberation from biological constraints. The other is the freedom of the wild animal to live one's life as it lives without the need to become somebody else. We should therefore differentiate between therapy and enhancement. If we consider human enhancement a therapy, then all have the right to it and are entitled to have it. Our current condition becomes a disease. However, the answer need not be so straightforward. It depends on our attitude to life and everybody should answer it for himself. There are also alternative stories of what it means to be a human. It may be the case that our inability to be is just the result of the lost ability to be. The disability from the perspective of transhumanists may be considered an ability from the perspective of our current life, ability to have emotions, to live a short life, to deal with one's limits.

Transhumanists are healers of humanity (Hauskeller 2016, p. 122) as the current condition of man is considered a disease that needs to be cured. One of the diseases is the uneven distribution of talents and abilities in society caused by different genetic information, but our equality is also hampered by ageing, diseases or death. We should not forget that differences also have their value and make life interesting and rich. A certain degree of equality may be desirable; sameness is however not the best solution (Hauskeller 2016, p. 154). Another reason for not providing everybody with the same abilities in order to secure them equal chances in life is that we do not know which abilities are good and under what circumstances. Being straightforward may lead to breaking rules and lead the person to prison, but also may lead to success as the person will not have any barriers to be creative and break any standard and conventional way of behaviour. The idea of identical starting position is based on the myth that life is a competition where only the strongest survive. But life is not a race.

Even if we agree that some improvement of man's abilities is necessary and needed, the provision of its equal distribution would be difficult. Improving human abilities would not be available for everyone. That is why F. Fukuyama (2009) claims that the first victim of transhumanism will be equality. We all share a belief that humans have an equal value and the differences between men are only accidental. After changing human nature by modifying human essence the more advanced may become superior over those left behind. It may happen that unmodified humans would not be able to compete with the developed ones. People will be forced to enhance their potential in the future transhumanist society as the unenhanced ones would not be able to find their place in society and will be similar to the illiterate ones. In the new society learning and enhanced abilities will be necessary. The respect for human dignity will be harmed. In addition to that, Fukuyama (2009) says transhumanists determine what is good and bad in man. As a result, they may deprive man of some abilities that may be useful in some contexts or that are interconnected with other abilities and make a meaningful whole. Developing more powerful robots may in spite of their inability to perceive intentionally lead to human extinction, as man would not be able to compete with such machines. In some respects, transhumanism is similar to eugenics.

A barrier that stands in the way of the full autonomy is also the relation to the other. It makes us dependent on the object of love and other human emotions to other people. Here the limited character of transhumanism is proven as the other is not considered an option for development, definition of who I am, but a barrier. To be completely autonomous requires complete separation from other people which can be done either by reduction of their autonomy or by creating a world where I could do anything without the collaboration and assistance of others. However, it is difficult to imagine machines could ever love us. At least they could behave as if they loved us; they could maximally imitate a relationship and react to our activities, but their inner intentional relationship of love is not accessible to them and we still do not know how to provide them with it. Love is a double-sided relationship. If one part of the relationship (the machine) is not in this relationship, love is not fulfilled and remains potential. Love consists in some subjective feelings towards the other and machines cannot have such feelings. Love does not consist merely in a behavior; the lover may express caring or unconcerned behaviour and be in love. It is true, the feelings are expressed in behaviour, but if we have a plausible alternative explanation for its behaviour we would not ascribe it any intentions, says Hauskeller (2016, p. 186). We expect that the lover will really love us and not just behave as if he loved us. The big difference between a human and a robot consists in the fact that they are built for a certain purpose including to love us, whereas humans do not have any simple purpose. We can buy a robot to satisfy our needs, but it is humiliating to buy a human to satisfy them. What we consider important in the human lover is the fact that he may change his behaviour and stop loving us. With the real robot the interruption of love will not be possible and if it happened it would have causal or arbitrary reasons, with no free component. The other thus serves as mirror reflecting ourselves and as a source of disturbance of one's stability and self-contentment.

Man's dissatisfaction with his place in the world is expressed in distrust in his abilities, senses and bodies. The development of robots is one of the answers to the allegedly difficult situation of man who has to work in order to improve his miserable living conditions. However, if robots replaced man in all his necessary work, man would not know what to do. He is so much bound to his biological life (as labour is in fact the result of man's imperfection and dependence on nature) that he would be able to respond to biological impulses only. The popularity of simple amusement proves this conclusion. How can a society of labourers without labour look like? asks H. Arendt (1958). As man is not happy with his environment, he is not happy with his body, either. Man's body is considered imperfect, weak, bound to nature and in need of improvement. As well as world is seen as a realm of necessity, so too is human body. That is why people work on its improvement and strive for immortality, enhancement of their abilities, etc. The new technical devices distract our memory and attention to the world as they need a lot of attention themselves and so hide the human character of the world in various ways (amusement, simplification, automation, etc.).

Regarding improvement of man's abilities, the situation is more complicated. It is important to know what abilities are improved and for which reason. We cannot e.g. believe in a never-ending progress because progress does not proceed without disruptions. Every new invention has its negative effects which cannot be neglected. The technical progress is not automatically positive. For example, creating an environment fully controlled by man conceals from man natural world with its risks and so does not train his natural abilities. The real existence and its fragility cannot be perfectly imitated in the virtual environment. Technology may cause damage to the world, to the social well-being, to ethical harms like loss of privacy, etc. Technical improvement of human bodies may have harmful effects and cause more negatives than benefits. Human body is a well-balanced system and we do not understand it completely. Genetic manipulation or organ transplantation can be an example. Improving technological abilities does not necessarily lead to improvement of human morality. We must be sure what our aim actually is. Technological progress can lead to global regress after all. The extent of the changes that transhumanists propose is very large. The moral problem related to that is men are playing gods and so may trespass the abilities they are able to control. Let us consider the Tower of Babylon, Golem, Frankenstein, etc. as some deterrent examples.

6 Conclusion

The idea of transhumanism is dangerous in this context as it puts us in a position where we could need artificial machines to think and speak for us, as we will not be able to do it ourselves. Politics in the sense of evaluating and discussing issues of the common world requires meaningful speech. The new conditions developed without understanding may create a world people would not be able to understand. The progress of globalization accelerated by ICTs has already shown incomprehensible consequences against which people protest. The political consequences of new technologies should be discussed. Big data also provide knowledge without understanding that may be useful, but difficult to deal with and to discuss. We are losing the sense of what we are doing with modern technologies available. Transhumanists have lost the respect for intentionality, bodily existence, freedom, involvement and other phenomena that characterize human existence in the world. In history the examples of the Tower of Babel, Golem, Frankenstein, etc. show how dangerous the actions of man playing the role of god can be.

Transhumanists enhance human abilities without clearly explaining the sense and reason of such actions. Man is a balanced system of various components and we still do not understand the balance properly. Increasing or strengthening some of the powers may destroy that balance.

Unintended consequences may be another harmful outcome of the transhumanists' efforts. Even with genetic manipulation, we fear the consequences for nature and with human body manipulation we may get similar results.

In order to accept and respect reality one needs stillness and protected environment. With our orientation on labour we do not know how to keep still; we know either activity or passive consumption of the activity's results. We are not happy with ourselves and that is why we seek for diversion and amusement in our free time. However, we rather need cultivated attention, respect for the surprises of the world (respect and not government of nature) and patience. We need to get rid of the eternal processual time of the world and understand our life as an interval between life and death. Such a time conception focuses on the present moment spanned between past and future instead of the processual concentration on the future. Such a life can change and interrupt the automatic series of events and constitute something new brought about by human activities. In such understanding of the world we would be less oriented on biological needs and their satisfaction, but would be more focused on creativity, which produces durable objective things into the world, and political space that evaluates, compares and praises the objects of the common world (Braun 2007).

Acknowledgements The paper was written within the project "Systems Approaches to Information Ethics" supported by the University of Economics in Prague and the Bertalanffy Centre Wien under the helpful leadership of Prof. W. Hofkirchner.

References

Arendt, H.: The Human Condition, 2nd edn. University of Chicago Press, Chicago, IL (1958)

Bostrom, N.: A history of transhumanist thought. J. Evol. Technol. **14**, 5, 1–25 (2005)

Braun, K.: Biopolitics and temporality in Arendt and Foucault. Time Soc. **16**, 5–23 (2007). https://doi.org/10.1177/0961463X07074099

Brentano, F.: Psychologie vom empirischen Standpunkt. Duncker & Humblot, Leipzig (1874)

Fukuyama, F.: Transhumanism. Foreign Policy. http://foreignpolicy.com/2009/10/23/transhumanism/ (2009). Accessed 20 Aug 2018

Hauskeller, M.: Mythologies of Transhumanism. Palgrave Macmillan, London (2016)

Hayles, N.K.: How We Became Post-Human. The University of Chicago Press, Chicago (1999)

Huxley, J.: Transhumanism. J. Humanist. Psychol. **8**, 73–76 (1968). https://doi.org/10.1177/002216786800800107

Jaspers, K.: Philosophy of Existence. University of Pennsylvania Press, Philadelphia (1971)

Kass, L.: Ageless bodies, happy souls. New Atlantis. **1**(1), 9–28 (2003)

Lewis, C.S.: The Abolition of Man. Macmillan, New York (1955)

McIntyre, R., Smith, D.W.: Theory of intentionality. In: McKenna, W.R., Mohanty, J.N. (eds.) Husserl's Phenomenology: A Textbook. University Press of America, Washington, DC (1989)
Mirandola, P.: On the Dignity of Man, On Being and the One, Heptaplus. Macmillan, New York (1985)
More, M.: The philosophy of transhumanism. In: More, M., Vita-More, N. (eds.) The transhumanist reader, pp. 3–17. Wiley-Blackwell, Chichester (2013)
Nietzsche, F.: Thus spoke Zarathustra. In: Schlechta, K. (ed.) Werke in drei Bänden. Hanser Verlag, Munich (1966)
Searle, J.R.: Intentionality and its place in nature. Dialectica. **38**(2–3), 87–99 (1984a). https://doi.org/10.1111/j.1746-8361.1984.tb01237.x
Searle, J.R.: Minds, Brains and Science. Harvard University Press, Cambridge, MA (1984b)
Smith, J.M.: Science and myth. In: Hull, D.L., Ruse, M. (eds.) The Philosophy of Biology, pp. 374–382. Oxford University Press, Oxford (1998)

Chapter 5
Elements of a Posthuman Philosophy of Art

Stefan Lorenz Sorgner

Abstract The chapter shows how Western art history has been dominated by dualistic ontological premises since dualistic thinking was created with Plato's philosophy in ancient Greece. With Darwin's research and Nietzsche's reflections, the dualistic Western cultural tradition has undergone a twist. Since then, various non-dualistic ways of thinking together with this artistic creations have been realized. In recent times, the various posthuman philosophies, among which critical post- and transhumanism are the most widely received approaches, have led to new artistic modes, and corresponding challenges. On the basis of these reflections, I develop elements of a posthuman philosophy of art, whereby a specific focus is given to bioart which deals with aspects of both critical post- and transhumanist reflections. Both approaches generate new ways of thinking about the world which imply a fundamental parading shift. Many posthuman art works fulfil the demands of a total work of art, but one without totalitarian implications, i.e. a non-totalitarian total work of art.

1 Introduction

I have become fascinated concerning grand narratives already during my teenage years. It was then, when I began to realize how widespread categorical dualistic ontologies are and that they can be found in various fields, levels and strata of culture and life. When I am talking about these kinds of dualities I am referring to distinctions like the one between good and evil, mind and body, culture and nature, the material and the immaterial or the organic and the inorganic. The examples I mentioned are an arbitrary choice and several others could be mentioned, too. One could wonder what is problematic with these distinctions, as we are using them every

S. L. Sorgner (✉)
John Cabot University, Rome, Italy

W. Hofkirchner, H.-J. Kreowski (eds.), *Transhumanism: The Proper Guide to a Posthuman Condition or a Dangerous Idea?*, Cognitive Technologies,
https://doi.org/10.1007/978-3-030-56546-6_5

day, and it is at least not immediately clear why employing them could be problematic.[1]

The problems, which I started to realize first when I was still a teenager, were the ones related to the distinction between the immaterial mind and the material body. If human beings consist of two such radically separate substances, how could it be possible that mind and body interact with each other? If two substances do not have anything in common, then any kind of interaction seems highly questionable (Sorgner 2007, p. 46).

The next thing I realized were the evaluations which were attributed to the two substances. The immaterial world was usually related to the good, stability and unity. The material world on the other hand was connected with evil, change and plurality (Sorgner 2010, pp. 193–211). This way of conceptualizing the world is related to the assumption that the good is something which is universally valid. The good stands for qualities connected with the notion of a good life. In this way of thinking, a good life can be described, and the description is universally valid for all human beings since anthropologically all human beings are identical insofar as they are all possessing an immaterial personal and rational soul, which is identical with their true human nature, and which separates human beings categorically from all other solely natural beings like apes, dolphins or elephants. This way of thinking can still be found in many social contexts, legal constitutions and moral laws.

Having reflected upon the question of duality and non-duality for a long time, only recently I managed to connect two insights which I have had for some time, without considering that there could be a connection between them. It concerns the thought that there is a relationship between the birth of dualistic thinking and dualistic media and that there is an intricate link between the coming about and dominance of Platonic thinking and the birth process of ancient tragedy, as both are rooted in a dualistic manner of grasping the world.

In the beginning of August 2013, just before attending the World Congress of Philosophy in Athens, the Spanish artist Jaime del Val and I were on the island of Aigina and decided to attend a performance of Euripides' "The Cyclops" in the theatre of Epidauros. It is the only complete satyr play which has survived. During the performance, when I was confronted with the architectural prerequisites which were brought about by the institutionalization of drama which took place during the sixth century BCE, I suddenly became aware of the dualities which emerged during the birth process of ancient Greek drama.

Originally, there were no theatre buildings, there was no stage and there were no spectators who were separated from the stage. Before the institutionalization of tragedy, there were only groups of human beings singing and dancing together without a rigid dualistic spatial separation between the actors and the audience. Various categorical dualities were introduced during the birth process of tragedy (Pickard-Cambridge 1927).

[1]Selected thoughts were integrated from a different short paper of mine (Sorgner 2016).

Firstly, there was the spatial separation between the audience and the actors. The audience had to remain seated within certain linear and circular fields which were separated from but also directed towards the circle or rather stage on which the actors were supposed to fulfil their tasks.

Secondly, a distinction between the chorus and the protagonists was introduced. On the one hand, there was the chorus, and the task of the chorus was to sing and dance together. On the other hand, there were the individual actors whose task was to recite their roles. Hence, the duality between audience and actors was amplified by further introducing the duality between protagonists and chorus. Thirdly, the dualistic architecture of the theatre was created which enforces these dualistic structures. All of these dualities were absent from the festivities which took place before the invention and institutionalization of the theatre which started with the Theatre of Dionysus in Athens during the sixth century BCE (MacDonald and Walton 2011).

I am not claiming that the institutionalization of tragedy which came along with the construction of the Theatre of Dionysus was the sole event during which dualistic media (here: dramatic theatre) came about. However, it seems plausible to claim that this event was a central stepping stone during the historical process of the birth of dualistic media.

The same can be observed in the realm of philosophy. Dualistic thinking in the Western tradition was strongly influenced by Plato's thinking during the fifth century BCE. But we can also find dualistic conceptions before Plato, for example in Zoroaster's thinking during the first half of the second millennium BCE. Still, Plato can be seen as one of the key figures responsible for introducing dualistic ontological categories into the Western cultural tradition.

In Plato's case, the dualism can be found between the realm of forms and the material world. Even though he introduced a dualism between human beings who possess rational souls on the one hand and animals who do not have such souls on the other hand, this separation was not yet as rigid as it became later on, because Plato also stresses that there are several types of souls—a vegetative, a sensitive and also a rational soul. Any type of soul or psyche is responsible for self-movement and hence for life. Whatever has a soul lives. Consequently, Plato has good reasons for attributing certain types of souls (but not a rational soul) to plants and animals, as both are capable of directed self-movement which is a reason for attributing a type of soul to them. Yet, Plato regards the rational soul to be solely present in human beings and argues that a rational soul is necessary to be able to enter the realm of forms and grasp the forms, to use language and communicate via language with one another.

The next central step during the development of dualistic ways of thinking occurs with the Stoics. Stoic philosophy upholds that there is a unified *logos* which encloses immaterial human souls. Animals were not regarded as possessing such immaterial souls according to Stoics. The main difference to Plato concerning the question of duality has to do with the idea of *humanitas*. Plato did not think that just because all human beings possess a rational soul they also ought to be treated equally well. He affirmed that there were human beings with gold, silver and others with iron in their souls (metaphorically speaking), and their social rank depends on the type of metal one has in one's soul. Stoic philosophers, on the other hand, introduce the notion of *humanitas* which was linked to the equal evaluation of all human beings. This notion was transformed by Cicero into the concept of dignity which all human beings were supposed to have in an equal manner because they all possess a rational soul and belong to the human species. Even though it was obvious to Cicero that human beings differ with respect to their talents and capacities, he also acknowledges that human beings ought to be treated well solely for being a member of the human species. Stoic philosophers or Cicero did not yet develop an egalitarian society in the modern sense, yet, this transformation in the understanding of human beings did also have some practical implications, e.g. concerning the treatment of slaves in their society, as they were gaining a higher social recognition during this period of times.

A third crucial step in the development of dualistic thinking took place with Descartes and his philosophical outlook. In contrast to the ancient thinkers within the Platonic tradition who acknowledge that there are a variety of different souls, Descartes introduced dualism on an even more rigid level by introducing the distinction between *res extensa* and *res cogitans*. According to Descartes, human beings belong to both types of substances while animals and all other solely natural objects belong to the realm of *res extensa* only.

This kind of rigid dualistic thinking was developed further within the Kantian approach where we can find the same ontological distinction as in Descartes' philosophy. However, Kant focused more on the ethical relevance and implications of this dualistic understanding and developed a complex ethics and political philosophy which still serves as the inspiration for the basis of the German foundational law. Due to this influence it follows that it is still legally forbidden to treat other persons solely as a means which presupposes a radically dualistic distinction between objects and subjects. Furthermore, this influence is the reason why according to the German foundational law only human beings possess dignity, but animals and all other solely natural entities are supposed to be treated like things. This legal distinction presupposes a highly problematic categorically dualistic ontological separation which was already fundamental in Descartes' philosophy.

Here it might be interesting to note that all the categorically dualistic ontologies just mentioned do not directly have racist or sexist implications, even though it cannot be doubted that such associations were culturally established in connection with such ontology. Still, the philosophies just mentioned do not refer to and justify that white, heterosexual, rich men represent a cultural ideal of perfection. Still, it is the case and it cannot be doubted that culturally the immediate connection between white, heterosexual, rich men and an immaterial rationality was established. On a philosophical level, the shift from dualistic to a non-dualistic ontology was far more important than any later cultural association which was connected to this categorically dualistic ontology. Philosophically all of the thinkers mentioned held that women possess rationality. They also affirmed that a human cyborg consists of an entity with a rational soul and a material body. It was this view which was challenged from the nineteenth century onwards, in part by the great variety of posthuman philosophers. I use the term posthuman philosophers to refer both to post- and to transhumanists, because the notion of the posthuman comes up in both traditions, even though a different meaning is associated with this word within these traditions.

Yet, both traditions doubt that a categorically dualistic ontology is an appropriate anthropology (Ranisch and Sorgner 2014).

After Kant, Nietzsche moved beyond the dualistic history of Western philosophy and the impact on and all the consequences of his approach have yet to be grasped by scholars, thinkers and philosophers today. However, Nietzsche, together with Wagner, Darwin and Freud, has initialized a cultural move towards a non-dualistic way of thinking. Consequently, it is possible to stress that with this cultural shift, humanism in its traditional form is coming to an end. Here, I understand humanism as a worldview which is founded upon a categorically dualistic ontology. This understanding is in tune with the etymology of the word "humanism" which comes from the Latin "humanitas". This concept was central for Stoic thinking, and it implies a categorically dualistic ontology.

Given that the aforementioned reflections concerning the development of dualistic thinking are plausible, it needs to be realized that the development of Plato's philosophy has most probably been the central cornerstone for the foundation of Western culture as a dualistic culture. Sloterdijk (1999), who identifies the beginning of humanism with the age of Stoic philosophy, and Hassan (1977), who stresses the close connection of the beginning of the enlightenment with the beginning of humanism, are correct in claiming that strong versions of dualisms can be found in the philosophies of the Stoics and of Descartes. However, it would certainly be highly implausible to disrespect the central importance of Plato's philosophy for this development.

As a consequence of the breaking together of humanism, several cultural movements have emerged that move beyond categorically dualistic ontologies today. Consequently, it seems appropriate to claim that we are moving beyond humanism into the age of the posthuman, whereby the posthuman as an open metaphor stands for a great variety of beyond humanism movements like post- (Hassan 1977), meta- (Del Val and Sorgner 2011) and transhumanism (Huxley 1951) in which the word "posthuman" comes up and which have in common that they doubt the ontological foundation of humanism. Still, it needs to be stressed that the goals, pedigrees and methodologies of the various movements differ significantly.

2 Non-duality, Technology and Posthuman Works of Art

Non-duality has already been a central feature of many postmodern works of art, as Welsch correctly noted (Welsch 2007, p. 110). What was missing in postmodernism was the focus on technology which is one of the central features of a posthuman work of art. Maybe, by means of the following reflections we will be able to get a clearer understanding of these.

I begin with Kevin Warwick's works because they dissolve the categorical dualities of mind and body and organic body and inorganic things. He is not an artist or a media maker, but an engineer. However, many of his works reveal excellently the shift towards non-dualistic media and hint at central philosophical issues which go along with it. This shift is not only related towards a dissolution of non-dualistic categorical ontologies, but it also goes along with the realization of the importance of the *cyborg*. "I would rather be a cyborg than a goddess" (Haraway 2004, p. 39). Technologies and cyborgs are particularly important, as they represent best the dissolution of categorical ontological dualities which have been occurring recently. A cyborg is not just a hybrid in the sense that it is a mere mixture between two categorically separate substances (a material and an organic one), but it is a synthesis of elements which merely seem categorically distinct, but are not so. "We have always been posthuman" (Hayles 1999, p. 291). Even a pacemaker has mental elements. A thing is not just a thing. If a policewoman takes away my smartphone, she violates my right to life and physical integrity.

Here I will only focus on one of Warwick's many developments, yet it represents a central realization which goes along with the posthuman turn. Kevin Warwick established a brain computer interface, a neural implant technology, which allowed him to connect his nervous system via a computer to the Internet while physically being in New York. The signals he was sending out were sent via the Internet to a mechanical arm in his laboratory at the University of Reading in the UK. Despite the distance, he managed to grab or touch objects with this connected mechanical arm. When grabbing or touching an object, the sensors in the fingertips of the mechanical arm sent the sensory input back via a computer and the Internet to the brain and nervous system of Kevin Warwick sitting in a room at Columbia University in New York. While one could argue that this was just a—distant—mechanical arm, the ability to feel the mechanical fingertips touching an object blurred the boundaries between what is a mechanical arm and his arm. He did not try this experiment with apes first but directly took the risk of establishing this feedback mechanism with his brain. As no one else had ever done this before him, there was the risk of his brain being permanently damaged because it was uncertain what exactly would happen to him. However, the experiment was successful providing us with sound reasons that the rigid categorical separation between mind and body or the organic and the inorganic we used to believe in does not hold.

Concerning artworks, Jaime del Val's meta-body project is probably one of the most promising ways of moving beyond a dualist media. Here I will focus on the "Pangender Cyborg"—Metaformance by Jaime del Val which can be seen as a central preparatory work for the meta-body project. It is important that it is a metaformance, and not a performance, as performances presuppose the categorical distinction between audience and performer which a metaformance attempts to transcend. For this metaformance of his, he developed a device that consists of several cameras placed on different parts of his body. Then, a projector attached in front of his chest projects the images from the cameras into the space in front of him. Additionally, loudspeakers on his back amplify the sounds he is making, which again amplify and support the affects of his becoming during the metaformance. In particular the issue of meta-sex is closely related to the interplay and interrelatedness of sounds, post-anatomical perspectives and his movements of amorphous becomings. Jaime del Val's metaformances challenge and problematize traditional dualistic ways of thinking in philosophy on several levels.

Firstly, his metaformances reveal the non-duality of ontology. Christian and Kantian traditions distinguish between objects and subjects and hold that these are categorically distinct and separated from each other. Del Val's metaformances dissolve these distinctions revealing the permanent relationships in which we are embedded and that we are merely "becomings" in this world. It is not even possible to adequately express this thought propositionally, as our grammar affirms a different type of ontology. The cameras capture small aspects of his body, an unusual perspective of his hand and thumb, whereby the perceiver does not immediately realize what is being depicted, in other words a post-anatomical perspective of this part of his body, and project these images on the walls around him. However, by perceiving whatever one is surrounded by, oneself gets affected. Hence, unusual details of his body firstly get amplified and magnified and any slow movements of the body parts appear faster in the magnified projected image in front of him. This, in turn, has an impact on him: he is interacting with himself or better with various post-anatomical perspectives of his body. Slowly he is moving forward, and via the projections of his movements, he is altering his future movements. There is a permanent interaction with him taking place, as he is part of permanent processes of amorphous becomings. This metaformance is a strong criticism of the rigid subject and object distinction of dualistic ontologies.

Secondly, I have already briefly mentioned that the perspectives of the cameras are unusual ones as they do not divide the body into traditional anatomical parts. Thereby, the contingency of anatomical classifications is being revealed. It is possible to classify the body in many different modes. Hence, the post-anatomical perspectives, which are part of this metaformance breakup encrusted linguistic structures, reveal the contingency of categories and, thereby, open new fields of becoming.

Thirdly, the post-anatomical perspectives which are being enhanced and supported by sounds produced by his meta-body challenge the traditional dualistic conception of sexuality during this metaformance. Dualistic concepts of sexuality reduce sexual relationship to the genitals which are being classified in a binary fashion. The meta-body, however, confronts these binary distinctions with the help of post-anatomical perspectives and the corresponding sounds. Thus, it can become clear that sexuality can be present in an unusual way of perceiving an ear, a shoulder or a leg, in the way we approach a foot, in a new sound, a scream or a shout we are being confronted with. There is an enormous multiplicity of possible relationships which can be grasped as sexual and which can be connected to intense feelings of gratification. Hence, this metaformance also enables us to move beyond a binary concept of sexuality towards a movement which can be helplessly referred to as meta-sex.

3 Elements of a Posthuman Philosophy of Art

Even though, there are good reasons for claiming that the birth of posthuman art is currently occurring, it is still far from clear which notion of "art" is being employed which is characteristic of posthuman aesthetics. It is clear that an avant-garde aesthetics is not suitable for explaining why Alba the fluorescent rabbit should count as a work of art. On the basis of Frankfurt school aesthetics, it might be more appropriate to claim that Alba is an industrial advertisement tool for making genetic modifications look intriguing, but definitely not a work of art. Yet, Avant-garde aesthetics is the strongest aesthetic tradition of the twentieth century, at least within the continental philosophical tradition. The challenges of accepting posthuman works of art as proper art can only be fully grasped against this cultural backdrop.

According to Adorno, the authenticity of a work of art is important for dialectically making the recipient aware of non-autonomous social structures within a social system (Adorno 1970). Hence by being autonomous, authentic works of art can have a heteronomous effect upon recipients. As numerical beauty has a long tradition and rigid implications, it cannot be part of an autonomous work of art, and has to be avoided. According to Adorno, there has to be conflict between being authentic while one is at the same time drawing upon traditional beautiful forms for one's own work of art which was the case for Stravinsky.

Consequently, neither a musical piece like Koyaanisqatsi by Philip Glass nor Alba by Eduardo Kac could count as art. Both would simply be too beautiful for being regarded as proper autonomous art. Adorno would accuse both makers as being inauthentic, as they were influenced too strongly by the culture which surrounds them in which you find such traditional numerical relationships which are being associated with beauty. However, an autonomous artist by being authentic could never find these objective numerical relationships within herself or himself. Consequently, it would have to be the case that these media makers were strongly influenced by what one does when creating these pieces, rather than being authentic. All they managed to do was to create Kitsch, sentimental design objects for the masses.

I understand his line of reasoning, but I disagree with it. On the contrary, it needs to be noted that the Frankfurt School is still stuck in a categorically dualistic ontology itself and can therefore still be seen as a representative of the Platonic-Christian dualistic culture without realizing that this is the case. Habermas' anthropology exemplifies the issue in question explicitly. He refers to his anthropology as a weak naturalistic one (Habermas 2004, pp. 876–877). However, when considering his reflections in detail, it becomes clear what he means by this, namely that even though he accepts that most aspects of human beings are empirically accessible, this is not the case for all of them. What does this imply? It implies that some human aspects in principle cannot be analysed empirically, which can only be sensibly explained by reference to some kind of immaterial rational soul. However, it is this kind of thinking which has been overcome by posthuman approaches.

Still, the question needs to be tackled, how to evaluate the above-mentioned musical piece Koyaanisqatsi and the living sculpture Alba from an aesthetic point of view, and how should we deal with all the other posthuman works of art. Should they even count as works of art?

If you take Darwin and his theory of evolution seriously, and in particular for transhumanists this is central for their way of thinking, the question arises concerning the relevance of art from a naturalist and an evolutionary perspective. What is the evolutionary meaning of listening to a symphony? What is the relevance of there being an aesthetic attitude which includes disinterestedness, given that we are merely psychophysiological organisms which are embedded in a permanent struggle for existence in which only the fittest ones survive? Should not art be connected to evolutionary processes and natural selection, too? Should it not be the case that our capacity to perceive beauty is also related to these processes? Why should we develop a disinterested capacity for aesthetic perceptions, if we are permanently confronted with the possibility of death, both as a species and as an individual?

Due to these questions, evolutionary philosophies of art developed explanations for us having gained a capacity to perceive beauty, and what beauty means, if we analyse it scientifically. By means of psychological tests, it was realized that the closer a face gets towards the numerical ratio of the golden mean, the more people regard it as beautiful (Langlois and Roggman 1990). So it seems to be that there is some validity concerning the Pythagorean-Platonic account of beauty, at least with respect to the numerical analysis of these concepts, in part the golden mean and harmonious numerical ratios like 1/2, 2/3, ¾. Evolutionary accounts of art support the analysis of the Pythagorean-Platonic tradition that some types of beauty can be analysed by reference to certain numerical ratios. However, the reason they suggested were different from the ones put forward by the Pythagoreans and Plato. While the latter referred to a separate realm of ideal forms of numerical perfection, the former explain the validity of these ratios by reference to a type of strength which increases our fitness so that we increase the likelihood of being appropriate for the natural selection procedure. Fitter human beings have more sonorous voices in which you find many overtones. Consequently, humans developed a capacity to identify these ratios independently and we started to identify harmonious ratios with something desirable, i.e. beauty. The same logic can be used with respect to the golden mean. Healthy and strong human beings have a physiological presence in which you often find the ratio of the golden mean. Hence, this numerical ratio began to being associated with something desirable, i.e. beauty.

However, simply by considering some specific cases, one can detect the limits of this theory, e.g. in the case of peacocks (Welsch 2016, pp. 65–65). The tail of the peacock is beautiful. However, it is long and when it gets opened up, it gets much more difficult for the peacock to get through the woods. So a beautiful tail does make survival more difficult, as it renders difficult the task of getting through the forest. Still, having such a tail is clearly in the peacock's interest, as it increases the likelihood of getting a partner and having offspring, i.e. it increases the likelihood of being successful in the sexual selection procedure. So there even seems to be a tension between natural and sexual selection. In the initial explanations, we had reasons for holding that beautiful organisms are good at passing the natural selection procedure. With the peacock's tail, we have an example where this is not the case. However, a beautiful tail is still in the peacock's interest, as it increases the likelihood of winning the sexual selection procedure. In both instances we have types of beauty. Both have a functional purpose, a purpose which can be explained by reference to evolutionary processes. Obviously, a lot more could be said on this issue. I am particularly drawing upon the fascinating recent work by Wolfgang Welsch (2016) on this topic.

As the second example shows, even in the animal kingdom there can be a perception of beauty which is not immediately connected to the aspect of natural selection. Maybe, these more developed cultural types of beauty can have come about when the issue of mere survival has become less relevant. Most members of the species manage to survive, but there still arises the need to distinguish oneself so that the best possible life can be realized. Here any type of social distinction can promote our recognition by others and can demonstrate and affirm your power status. In this context, the concept of power is not meant in a political sense only, but in wide sense so that it can apply to a great variety of social and cultural phenomena. The beautiful tail of the peacock is a sign of power. I can survive, even if I have a tail which would normally hinder me doing so. This is one explanation for the coming about of other more complex cultural concepts of beauty. It is a sign of my strength that I can dedicate myself to seemingly meaningless or maybe even better functionless activities. I am strong, I am free, I can do whatever I want to do. It is this understanding of beauty which becomes relevant in posthuman aesthetics.

Koyaanisqatsi is beautiful, and it is a posthuman work of art. It can be beautiful, as there is no ultimate reason which should prevent us from drawing upon and employing a numerical understanding of beauty in works of art. This does not imply that a work of art ought to be beautiful. At the same time, the relevance of beauty can be analysed on the basis of reflections about power structures. Nietzsche and Foucault have taught us well how to do so. Here, beauty and power get reconnected. Some reflections from an ancient philosopher go into a similar direction. Lucretius explains that after the techniques have been developed which are helpful for human survival, the non-necessary arts got developed which are a prerequisite for a pleasurable way of structuring one's leisure times. Music in Lucretius gets identified with a means of achieving the highest kinds of pleasure (Rumpf 2010, pp. 217–232). An implicit connotation of this thought is that leisure, music and pleasure can only be enjoyed by powerful ones, the aristocracy and the ones who do not have to worry about simply staying alive. Again, you find the connection between the arts, leisure (otium) and power. I am financially so well off that I can reflect upon the meaning of the world which does not serve any practical purpose. The liberal arts are the arts which are suitable for human beings who are free from the necessity to work.

This analysis would also be in tune with an evolutionary account of art. Yet, in contrast to a simple-minded reductionist evolutionary philosophy of art which claims that beauty always has to get reduced to a simple process of natural selection, here, we go beyond this simple-minded theory of art and stress that an account which takes evolution seriously can still recognize that there are many concepts of beauty which are entirely culture dependent. Or in other words, a simple-minded claim that all arts should incorporate the golden mean is not necessarily a claim which needs to be made on the basis of an evolutionary philosophy of art. However, the capacity to perceive harmonious beauty has become part of the human psychophysiology which means that it can be part of an authentic work of art to include beautiful aspect. In contrast to Adorno, we can even stress that by transcending the prohibition of beauty in artworks, we even increase plurality and diversity within the art world, and plurality is a quality which is also being affirmed from an evolutionary perspective, as plurality increases the likelihood of passing the natural selection procedure, which depends upon the fitness of an organism. The more different organisms we have, the more likely it is that some of them fit their environment such that they can survive. This is an insight which we can also apply to the aesthetic evaluation of the works of Sven Helbig, or to some works which fall into the category of bioart like Kac's fluorescent bunny Alba.

4 Sven Helbig

Sven Helbig is among the most fascinating living composers. His stage work is attributable to the tradition of the Gesamtkunstwerk. Incidentally, in his most important ancestor, Richard Wagner, there are some anticipations to posthuman positions. For example, the gods of his opera "Rheingold" depend on the eating of Freia's golden apples in order to retain their divine qualities of strength and youth. There are, therefore, structural analogies between these gods and the posthumans, as described by transhumanists, for which posthumans represent a further developed form of human existence. However, references to posthumanist positions in Wagner's work are also to be found. Here, for example, the language that occurs in his musical texts is to be mentioned. This does not only sound unusual for us today, but also does not represent the everyday language of the nineteenth century. Wagner was aware of the fact that the words always convey ideological contents and that a rigid subject–object distinction is closely linked to the German language. It itself affirmed immanent, naturalistic and evolutionary thinking. In order to avoid the dualistic implications of German grammar, he developed his own personal, metaphorical language, which he used within his musical dramas. These two examples show that Wagner's works contain post- as well as transhumanist elements. The general social, political and ethical orientation of his work, however, involves an orientation that is in conflict with posthuman thinking, and which involves numerous potentially problematic implications.

This assessment does not apply to Helbig's work. His music drama "From the Noise of the World or the Revelation of Thomas Müntzer" successfully avoids the potentially totalitarian connotations of the concept Gesamtkunstwerk. It nevertheless addresses ontological, ethical and political questions. However, it does not remain within the mythic realm, but always refers to current bioethical and religious challenges. What is the relationship between religious and political foundations? What moral assessment is appropriate for ethical questions at the beginning of life? Should utopias play a role in everyday political decisions? Within the final scene of the work “From the Noise of the World”, the demons emphasize that we are doomed when we follow a utopia, a general order and strong ideas. In this way, the radical plurality of good is stressed, and so is the fact that the ethical nihilism of our time is an achievement and not a loss. These are posthuman insights. Thanks to the use of the latest technologies, innovative media and an accessible musical language, Helbig avoids that the reception of this work is limited to a specialized audience.

Similar considerations are also to be made with regard to his concerto “Pocket Symphonies Electronica”. In doing so, he reverts to orchestral recordings of his own music, plays live to her as an instrumentalist and is at the same time responsible for the appropriate mixture and thus also assumes the role of a DJ. The separation between live music and recorded music is thereby subverted as much as that between serious and popular music, or the distinction between the composer and the musician, as was also the case in the ancient theatre. His instrumental music, too, is thus an inspiring plea for plurality and the softening of rigid traditional categories.

5 Bioart

Eduardo Kac coined the term bioart and many of his bioart works can serve as icons of posthuman artworks, whereby Alba and Edunia, a genetically engineered flower that is a hybrid of Kac and Petunia, might be the prime examples. Both artworks are innovative with respect to the material used, as traditionally works of art were seen as objects; his works, however, are living organisms which represents a paradigm shift in the history of art. It is a paradigm shift which concerning its relevance and radicalness can be compared to the Fountain by Marcel Duchamp or the Brillo Boxes by Andy Warhol. Which works of art can actually be categorized as bioart works is still subject to debate (Miah 2014).

The most obvious option is that a bioart work can be defined as an artwork which needs to be living organisms like Alba. If this is the case, then one could still wonder whether formerly living organisms should also be included, e.g. Damien Hirst’s famous shark sculpture. In addition, it might also be possible to regard Stelarc’s[2] ear or Orlan’s body modification processes as bioart works, as they fulfil the above criterion. An additional artistic realm concerns AI. Here, artists like the group

[2]There is a lot to be said concerning the relevance of Stelarc’s works for the posthuman tradition. His works bear transhumanist as well as posthumanist connotations. Concerning ethics, his performances are more closely connected to critical posthumanism than to transhumanism. Yet, all of the great variety of philosophical challenges his works raise are related to the posthuman turn. There ought to be a separate study concerning the manifold implications of his performances concerning the most fundamental aspects of posthuman philosophies.

Magenta[3] have created an AI as an artificial artist/artwork which itself is creative and creates further artworks. Is not creativity a sign of aliveness? Maybe, it is sufficient for an artwork to deal with the technological challenges of our times to count as a work of bioart. In this case, even a sculpture by Patricia Piccini could count as work of bioart, e.g. her work "Still Life with Stem Cells". Even though, the meaning of the notion bioart needs to be clarified further, it is clear that when we discuss posthuman artworks, the realm of bioart needs to be given special attention.

A lot more could be said concerning the relationship between bioart, robotic art, body art, performance art and how these disciplines relate to trans- and posthumanism. However, the main focus of my argument here is the underlying philosophical aesthetics of posthuman works of art. This is the reason why a detailed conceptual clarification concerning the various artistic disciplines related to the posthuman arts is not my main focus in this paper.

6 Total Works of Art

Besides the aspect of beauty, there is an additional aesthetic phenomenon which can be found in the artworks of Helbig or Kac, namely that there is a tendency to embrace totality, as the posthuman turn goes along with a radical paradigm shift. The Western cultural tradition was founded upon dualistic thinking and the relicts of this thinking are still culturally dominant in many realms and circles. Posthuman artworks represent suggestions concerning a new understanding of the world, which could be appropriate after the posthuman turn has taken place. However, in contrast to traditional total works of art, these do not regard their own suggestions as true ones which claim universal validity. It is this element which distinguishes them from Wagner's total work of art concept which has highly problematic totalitarian implications. (Some) Posthuman works of art can be characterized as non-totalitarian total works of art. In order to grasp the relevance of this phenomenon, we need to take a closer look at the notion of *techne*.

The ancient Greek notion of *techne* stood both for art and for technology. As a consequence of the humanist separation of mind and body, whereby the mind got connected with a non-empirical realm, whereas the body was connected to the sensual realm, art and technology got separated, too. Art became the sensual representation of the non-empirical (poiesis), whereas technology was merely a means for realizing immanent goals (praxis). With Helbig's music drama or Kac's bioart, the realms of art and technology got reunited, which corresponds to a wider cultural development which has occurred since Darwin and Nietzsche. With this cultural reunification of art and technology, the coming about of a non-dualistic, relational and a naturalistic, evolutionary way of thinking, i.e. several versions of

[3]http://europe.newsweek.com/can-artificially-intelligent-computer-make-art-462847?rm=eu Accessed 28 Oct 2020.

ontologies of becoming, occurs, too. However, it must be noted that a proper understanding of such ontologies does not claim their own overall superiority, as such strong claims concerning truth would demand a static ontological realm which are not consistent with ontologies of becoming.

Works of bioart represent total works of art, because of this reunification of art and technology. It is non-totalitarian, because it does not claim to be the only way art ought to be done. Jaime del Val's metaformances represent the rebirth of non-dualistic media, which is a direct consequence of the birth and death process which had occurred between 2500 BCT, and more recent developments. It is a total work of art, because it embraces and uses all facets of life; even the traditional audience gets included, which is very much in the spirit of pre-theatre dramatic works. Sven Helbig produces total works of art insofar that he dissolves the categorical distinctions between human beings and machines, and between composer, performer, DJ and improviser. These works do not demand any ultimate superiority, but represent different aspects of the birth of non-dualist media.

7 Posthuman Art and Non-totalitarian Works of Art

In the cases of Sven Helbig, Jaime del Val and Eduardo Kac, we can see elements which are characteristic of a total work of art, a Wagnerian Gesamtkunstwerk, as many of their works capture a totality of human experiences or use a totality of artistic means. Both definitions can be paradigmatic for a total work of art. These works challenge the aesthetic prohibition of total artistic structures by aesthetic theories such as Adorno's. Yet, these posthuman art works do not imply new totalitarian structures, but they increase plurality. This is the main difference between Wagner's total work of art and posthuman aesthetics which is characterized by being non-totalitarian total works of art, e.g. some of Eduardo Kac's works of bioart, Jaime del Val's metaformances as well as Sven Helbig's musical works. What is characteristic for all of them is that they neither stress their own superiority, nor do they claim a universal validity, but they merely represent a further offer.

Adorno's aesthetics demands art works which are dedicated to a permanently more intellectual audience. These posthuman artworks are inclusive without stopping to be innovative. Plurality gets promoted by including non-totalitarian total works of art in the spectrum of the contemporary art world. I am not claiming that all posthuman artworks are non-totalitarian total works of art, but it seems that there was a renaissance of the Gesamtkunstwerks tradition with the event of the posthuman turn, as this turn goes along with radical critique of the previously dominant Western culture which is founded on categorically dualistic ontologies and what these artworks do is to present alternative suggestions, new ways of perceiving and innovative sensual experiences.

8 Conclusion

In the aforementioned reflections, the birth and death of dualistic media has been analysed. It came out that there have been several versions of dualistic media in the past 200 years, e.g. both postmodern and posthuman works affirm versions of non-duality, whereby the latter in contrast to the former ones focus on technology. A central shift which has occurred together with the posthuman turn was the focus on an evolutionary way of thinking, which made it necessary to reflect upon the evolutionary meaning or art. Thereby, it becomes clear that it was possible to justify different understandings of beauty on this basis: (1) beauty as a natural fitness indicator; (2) beauty as an indicator of sexual attractiveness; and (3) beauty as a sign of power. As a consequence of these reflections, it has become clear that an evolutionary aesthetics does not imply the need of a Leni Riefenstahl aesthetics. Yet, it allows and explains the relevance of a formal concept of beauty which is based on the golden mean or a harmonic numerical ratio which does not imply that it ought to be used in works of art, but it implies that it must not be forbidden to use it in authentic works of art. Kac's Alba, the fluorescent rabbit, can count as a work of art on such a basis, even though its beauty would be a reason for it not being one on the basis of Adorno's aesthetics, as Adorno holds that such beauty cannot be part of an authentic work of art. By taking into consideration the theory of evolution, it is possible to explain why this does not have to be the case. It is possible to draw upon such an understanding of beauty and create an authentic work of art. In addition, it needs to be noted that by transcending the avant-garde prohibition of beauty, it is also possible to increase plurality and diversity within the spectrum of contemporary works of art. Both plurality and diversity again can be justified by reference to the theory of evolution, as these qualities increase the likelihood of human survival. An additional aspect of posthuman works of art is their widespread tendency of being total works of art by presenting alternative worldviews to the dominant Western understanding which is based on a categorically dualistic ontology, i.e. perfection consists in an immaterial reason which culturally gets associated with whiteness, maleness and masculinity. Posthuman suggestions criticize this notion of perfection and present alternative understandings. Furthermore, in contrast to Wagner's total work of art which claims superior validity and aims to become culturally dominant, these posthuman artworks are non-totalitarian total works of art, which do not regard their own suggestions as ultimately superior, and universally valid. By presenting their suggestions, they merely wish to undermine the dominant identities and present a great plurality of alternative suggestions.

References

Adorno, T.W.: Ästhetische Theorie. Suhrkamp, Frankfurt am Main (1970)
Del Val, J., Sorgner, S.L.: A metahumanist manifesto. Agonist. **4**(2), 1–4 (2011)
Habermas, J.: Freiheit und Determinismus. Deut. Z. Philos. **52**(6), 871–890 (2004)

Haraway, D.: A manifesto for cyborgs. In: Haraway, D. (ed.) The Haraway Reader, pp. 7–46. Routledge, London (2004)

Hassan, I.: Prometheus as performer—toward a posthumanist culture? Ga. Rev. **31**(4), 830–850 (1977)

Hayles, N.K.: How We Became Posthuman—Virtual Bodies in Cybernetics, Literature, and Informatics. The University of Chicago Press, Chicago (1999)

Huxley, J.: Knowledge, morality, and destiny. The William Alanson White Memorial Lectures, third series. Psychiatry. **14**(2), 127–151 (1951)

Langlois, J.H., Roggman, L.A.: Attractive faces are only average. Psychol. Sci. **1**, 115–121 (1990)

MacDonald, M., Walton, J.M. (eds.): The Cambridge Companion to Greek and Roman Theatre. Cambridge University Press, Cambridge (2011)

Miah, A.: Bioart. In: Ranisch, R., Sorgner, S.L. (eds.) Post- and Transhumanism. An Introduction, pp. 227–240. Peter Lang, Frankfurt am Main (2014)

Pickard-Cambridge, Arthur Wallace: Dithyramb, Tragedy and Comedy. Clarendon Press, Oxford (1927)

Ranisch, R., Sorgner, S.L.: Introducing post- and transhumanism. In: Ranisch, R., Sorgner, S.L. (eds.) Post- and Transhumanism. An Introduction, pp. 7–29. Peter Lang, Frankfurt am Main (2014)

Rumpf, L.: Lukrez. In: Sorgner, S.L., Schramm, M. (eds.) Musik in der antiken Philosophie: Eine Einführung, pp. 217–232. Königshausen und Neumann, Darmstadt (2010)

Sloterdijk, P.: Regeln für den Menschenpark. Ein Antwortschreiben zu Heideggers Brief über den Humanismus. Suhrkamp, Frankfurt am Main (1999)

Sorgner, S.L.: Metaphysics Without Truth—On the Importance of Consistency Within Nietzsche's philosophy. 2. bearbeitete Auflage. University of Marquette Press, Milwaukee, WI (2007)

Sorgner, S.L.: Menschenwürde nach Nietzsche: Die Geschichte eines Begriffs. WBG, Darmstadt (2010)

Sorgner, S.L.: The pedigree of dualistic and non-dualistic media: grasping extra-medial meanings. J. Relig. Film Media. **2**(1), 15–22 (2016)

Welsch, W.: Jenseits der Ästhetisierung. Ein anderer Rahmen für die Betrachtung der Werte. In: Jérôme, B. (ed.) Die Zukunft der Werte—Dialoge über das 21. Jahrhundert. From the French by Frank Sievers and Andreas Jandl, pp. 100–114. Suhrkamp, Frankfurt am Main (2007)

Welsch, W.: Ästhetische Erfahrung. Zeitgenössische Kunst zwischen Natur und Kultur. Fink, Paderborn (2016)

Part II
Military Aspects

Chapter 6
Transcending Natural Limitations: The Military–Industrial Complex and the Transhumanist Temptation

Christopher Coenen

Abstract For the military–industrial complex (MIC), transhumanism represents a temptation in various ways: its expectations for the future of technology imply there is a prospect of overcoming hitherto given barriers to the expansion of military power, particularly in terms of the linkage of humans and machines (and above all computers), while its characteristic concept of liberation, the aim of which is to transcend natural limitations, offers the possibility of weaving hopes for progress in the field of military research into an (at least superficially) emancipatory future narrative. In return, the MIC represents a temptation for transhumanism (as a societal movement) insofar as research and technology development projects that are (still) of little interest for civilian purposes can be driven ahead in a military setting. Ethical and political analyses of the relationships between military research and transhumanism may be enriched by historical and cultural perspectives. This is true for both, the fascinating pre- and early history of transhumanism before 1945 and the post-war history of this techno-visionary worldview. The transhumanism of our current times appears to have emerged from the intersections of military research, the new IT industry (for parts of which it has become an ersatz religion) and the counterculture of the 1970s.

1 Introduction

Nearly 10 years ago Jürgen Altmann described transhumanism accurately as "a strange mixture of old myths about humanity and technological euphoria, scientifically proven statements, more or less well-founded extrapolations, and seemingly dogmatic expectations", and drew attention to the fact that it was "increasingly" playing a role "in normal science, science funding and technology planning ...

C. Coenen (✉)
Institute for Technology Assessment and Systems Analysis (KIT-ITAS), Karlsruhe Institute of Technology (KIT), Karlsruhe, Germany
e-mail: christopher.coenen@kit.edu

W. Hofkirchner, H.-J. Kreowski (eds.), *Transhumanism: The Proper Guide to a Posthuman Condition or a Dangerous Idea?*, Cognitive Technologies,
https://doi.org/10.1007/978-3-030-56546-6_6

above all in the USA" (Altmann 2009, p. 51). In this connection, he also referred to the "NBIC initiative" in the USA (nano, bio, info, cogno = NBIC), which had held a series of events on "converging technologies" from 2001 on. These events featured various predictions about how key technologies and areas of research would grow together, so allowing increasing improvements of performance to be achieved by means of what is known as human enhancement (HE), a term that, to some, sounds rather like a euphemism for eugenics.

These technological visions were focused on, among other things, the expectation that the human being and technology would fuse together to a great extent. Apart from the "strengthening of the US military" and "vague promises", for example that "humanity could" become "a single, distributed, networked 'brain'", more medium-term visions of the future involving powerful brain–computer interfaces and the biotechnological modification of the human body were also formulated (Altmann 2009). The role transhumanism played in this initiative, on which US military research was one of the biggest influences, has been analysed a number of times (TAB 2008), while the many activities targeted at the propagation of HE have prompted calls for the political exploitation of futuristic visions to be interrogated critically (Nordmann 2007). It is in this context, in particular, that the concept of "vision assessment" has been further developed as part of technology assessment (see Ferrari et al. 2012). At the same time, it has been often emphasized that visionary futuristic discourse itself—and not only how the technologies in question are actually handled—should be approached in a responsible manner (Coenen 2011).

A declaration issued by the World Transhumanist Association (WTA) in the early 2000s still offers the best definition of transhumanist aims: "Humanity will be radically changed by technology in the future. We foresee the feasibility of redesigning the human condition, including such parameters as the inevitability of aging, limitations on human and artificial intellects, unchosen psychology, suffering, and our confinement to the planet earth" (Article 1, The Transhumanist Declaration 2002; quoted after Schneider 2009, p. 97). Viewed in this way, transhumanism essentially constitutes a project that is intended, above all, as a struggle against various, mostly natural, limitations and restrictions on the human.

With a view to the topic of the present paper—the interactions between military research and development (R&D) on the one hand, and this worldview on the other—it may therefore be noted at the outset that transhumanists propagate ideas about fighting nature (although this is in turn viewed as a natural task for humanity). Furthermore, if transhumanist technological visions were to be translated into reality, there would probably be a further explosion of military capabilities that would put considerable pressure on, or even sweep away, the rules that regulate warfare.

For the military–industrial complex (MIC)—which US President Dwight D. Eisenhower defined in his farewell address on 17 January1961 as a "conjunction of an immense military establishment with a large arms industry" and was decisively shaped by the (computer-based) technological revolution—transhumanism represents a temptation in various ways: its expectations for the future of technology imply there is a prospect of overcoming previous barriers to the expansion of military power, particularly in terms of the linkage of humans and machines (and above all computers), while its characteristic concept of liberation, the aim of which is to transcend natural limitations, offers the possibility of weaving hopes for progress in the field of military research into an (at least superficially) emancipatory future narrative. In return, the MIC represents a temptation for transhumanism (as a societal movement) insofar as R&D projects that are (still) of little interest for civilian purposes can be driven ahead in a military setting.

Does this mean the interactions between the MIC and transhumanism are evidence the two share an elective affinity? Before such interactions are looked at in rather greater detail in order to answer this question, transhumanism will initially be considered both generally and more specifically in terms of its significance for the IT industry.

2 Transhumanism

In view of the goals of transhumanism mentioned in the introduction, it is no surprise that (usually under other names—such as "extropianism"—and with a focus on the idea of "the Singularity"), it was initially tackled in studies by theologians, scholars of religion and other experts in the humanities, in particular studies of the history of ideas (for references to early discussions, see Coenen 2008). "The Singularity" is understood here as a caesura in the history of humanity: the moment at which artificial intelligence (AI) will be so far developed that a completely new era begins, one that will not be even remotely comprehensible before it dawns. However, that does not prevent transhumanist visionaries, such as the distinguished US inventor Ray Kurzweil, from making forecasts about this era (Kurzweil 2005). In essence, the concept of "the Singularity" corresponds to that of the "*intelligence explosion*", which was developed by Irving John Good in the first half of the 1960s (Good 1965). In our context, apart from the direct contribution to the history of transhumanist thinking Good made with this concept, it is of interest, among other things, that—like some of the most important developers of early futuristic transhumanist visions (John Desmond Bernal, for example)—he was deeply involved in the scientific work done to support the war against National Socialist Germany, and that his ideas about "*ultraintelligent machines*" constituted an outstanding example of the interactions between military research and AI research.

As a coherent ensemble of scientific and societal visions of the future, transhumanism is a product of the decades from 1870 to 1930 and took shape especially in Britain and Russia (and subsequently the Soviet Union). However, in Britain—whose intellectual history is of greater significance for contemporary transhumanism than the early-transhumanist thought of the tsarist empire and the

Soviet Union—it still did not have the function of a comprehensive worldview, but merely represented one element of progressive and, in particular, socialist ideological projects.[1] Some of the key British figures in this tradition were W. Winwood Reade, an author who travelled widely in Africa; the significant writer Herbert George (H. G.) Wells, a leading intellectual of his time; the outstanding scientist Julian Huxley (the brother of Aldous and first director general of UNESCO); John Burdon Sanderson (J. B. S.) Haldane; and Bernal.[2] In 1929, Bernal published what was—in the light of today's transhumanism—the most mature futuristic transhumanist vision of the period, the futurological essay *The World, The Flesh and The Devil* (1970). Almost all the core elements of transhumanist thinking are already found in this work, although of course ideas about AI, synthetic biology and neurotechnology could only be framed in highly speculative terms, if at all, in 1929. Even though their transhumanist visions do not appear central in Bernal's and Haldane's works—in comparison to their scientific research, their antifascist commitment, including its military aspects, and their other political activities (for instance promoting R&D that would deliver benefits for the whole of society or their pro-Soviet advocacy against investment in military might and war)—both let it be known during the second half of the twentieth century that these visions were of fundamental significance for their thinking.

There were then, insofar as can be ascertained, two main contexts within which early transhumanist thinking penetrated the (Western) MIC: firstly, military R&D in the service of the struggle against the fascist Axis powers, in which various scientists influenced by Bernal and Haldane were employed, and secondly, science fiction, which had been crucially influenced by the writings of Bernal, Haldane and Wells (see Parrinder 1995 and Slusser 2009) and achieved significant popularity in the English-speaking world from the mid-twentieth century on, especially among technoscientists. As early as the 1960s interactions were taking place between R&D on the one hand, and its fictional anticipation and reflection on the other, interactions that through to the present day have moulded discourse about the areas of science and technology that are of particular interest for transhumanism.

[1]By contrast, the early transhumanism developed in tsarist Russia by Nikolai Fyodorovich Fyodorov, who died in 1903, could be described as an all-encompassing worldview. This "cosmism", as it was known, has received greater attention in recent times in art and academic research. Bolsheviks like Leon Trotsky were aware of its ideological ambitions and consequently endeavoured—ultimately with success—to stifle its influence in the early Soviet Union (an influence that continued until after Lenin's death). On cosmism, see Groys and Hagemeister (2005).

[2]Furthermore, Bernal and Haldane were leading members of the circle of scientists politicized by communist or socialist ideas that came together in the Britain of the interwar period thanks to the significant influence of Soviet science policy. On this topic, see Werskey (2007) and Vogeler (1992). On these pioneers of transhumanist thinking, cf. Coenen (2013) and Coenen (2014a).

3 The Contemporary Transhumanist Milieu

Discourse about transhumanism is currently paying particular attention to the fact that it is a kind of ersatz religion for prominent representatives of the IT industry.[3] Here too, science fiction is a factor that should not be underestimated. For instance, Elon Musk (SpaceX, Tesla and other companies) and Mark Zuckerberg (Facebook), who both announced ambitious projects in the field of brain–computer interfaces (BCIs) during the first 6 months of 2017,[4] have been greatly inspired by a fictional world the Scottish science-fiction writer Iain Banks created for a series of novels about a post-shortage society called "*the Culture*" (see Cross 2017). In this world, power mainly resides with AI entities, which interact with what are for the most part human-like beings who enjoy numerous opportunities for "enhancement" in a generally anti-hierarchical, non-capitalist societal order.[5]

In the first half of the 2000s, Google's founders, Sergey Brin and Larry Page, also professed their belief in the core transhumanist visions of the creation of AI far superior to the human being and the connection of the human brain with computer technology, and they have repeatedly made similar statements in the 2010s.[6] Google and Alphabet's recent activities convey the impression that the translation of core transhumanist visions into reality (in AI and longevity research, for example) is genuinely part of the company's agenda. In this respect, the recruitment of Kurzweil, the most famous transhumanist, by Google, where he works as a "director of engineering" with a focus on AI (and not, as is frequently and erroneously claimed, as head of the AI department or even the whole engineering division), is often

[3]To give just one example: Irving John Good advised Stanley Kubrick while he was working on the famous science fiction film 2001: *A Space Odyssey* (1968), which features a paranoid supercomputer. The co-author of the script was the popular writer and futurologist Arthur C. Clarke who was also influential in space research. Clarke had worked in British military research during the Second World War and later wrote a novel informed by his wartime experiences, in which he described Bernal as one of the "Olympian entities" who were "rapidly changing the whole nature of warfare". See Clarke (1963).

[4]On this topic, cf. Millikan (2010), Levy (2011), McCracken and Grossman (2013), Coenen (2014b) and the references to other publications by journalists in Coenen (2014b). It is also an important topic of the ongoing transnational research project FUTUREBODY that analyses the rapidly expanding discourse on the merger of human corporeality with technology: https://www.itas.kit.edu/english/projects_coen18_futurebody.php.

[5]Elon Musk founded the firm Neuralink with what, for him, was a typically ambitious objective: the introduction of performance-enhancing BCI implants (including for non-therapeutic purposes) as early as the 2020s, in particular with the aim of making humanity fit for the coming age of AI. Facebook has formed a team with about 60 members to develop non-invasive BCI technology that will be suitable for everyday use.

[6]The turn some technology assessment experts have taken towards vision assessment has also strengthened in recent years—despite the relatively short time spans usually dealt with in policy advice—because the transhumanist worldview is evidently not merely being promoted by transhumanist organizations, a few academics (working in applied ethics, for example) and isolated political activities (such as the NBIC initiative), but also by key figures and companies in the IT industry that is so fundamental for our society today.

regarded as a symbolic act. Google also supported the establishment of the Singularity University, which was set up by Kurzweil with a partner at a NASA site, and defines itself as an institution dedicated to forging a global elite.

Various critical publications in German about transhumanism have attributed Google a key role in promoting it and, as it were, putting it into practice. The journalist Thomas Wagner warns that Google and other forces in Silicon Valley are in the process of installing a global "*robocracy*", and ensuring the human being becomes a "*discontinued model*" (Wagner 2015). The dream of "the Singularity", which is also harboured by some leading figures in the IT industry and can be understood as an extreme expression of the "*Californian ideology*" (Barbrook and Cameron 1996),[7] is interpreted by Wagner both as an ideology that is dangerous because it is so politically and societally influential and as a pseudo-religion. Markus Jansen—an author who has worked for the Gen-ethisches Netzwerk, a German civil society network that critically monitors developments in biotechnology—meditates at length on how transhumanism is redefining "*life*" in our "*age of global control*". He believes the "*merging of the human being with information technology*" aspired to by Google and other actors reveals "*a necrophilic, dead core*",[8] and is "*a signum of the latest totalitarian tendencies in a desolate global land*" (Jansen 2015, p. 290).

Other leading figures and significant corporations in the IT industry have also evinced a closeness to transhumanism, some of them ever since the 2000s: Bill Gates (Microsoft), who expects the spread of computer implants over the long term, has characterised Kurzweil as the best person at predicting the future of artificial intelligence, and he has been quoted by Kurzweil to the effect that he finds the creation of an optimistic new religion with a god-like AI desirable (Kurzweil 2005, p. 374f.). In 2008, Intel's Developer Forum was held under the motto "Countdown to Singularity", and Justin Rattner, the company's then Chief Technology Officer, devoted considerable time and attention to Kurzweil and his ideas.

From 2007 to 2012, the Machine Intelligence Research Institute at Berkeley headed by the controversial transhumanist Eliezer Yudkowsky ran what were known as "Singularity Summits" that were addressed not just by many transhumanists, but also by leading figures from the IT industry, such as Rattner, Peter Norvig (Google) and Dharmendra Modha (IBM). Apart from Yudkowsky, Kurzweil and the venture capitalist Peter Thiel were among the founders of this series of events. Thiel, who made a fortune with PayPal and invested in Facebook very early on, is a politically wayward "loony libertarian", who achieved a certain

[7]On this topic, cf. also McCray (2017).

[8]Since the first half of the 2000s, the radical French group Pièces et Main d'Oeuvre (piecesetmaindoeuvre.com) has been agitating against transhumanism, which it views as the avant-garde of a large-scale scientific, "necrotechnological", capitalist assault on humanity and life. Transhumanism appears to the group as a symptom and driver of the progress of inhumanity, and an enemy of the human. They believe anthropocide is being prepared in concrete, practical ways at technoscientific laboratories using biotechnology, information technology, nanotechnology and neurotechnology, a project that, with our yearning to remain human, we all have a duty to oppose. On this topic, see Pièces et Main d'Oeuvre (2015).

notoriety because he was the only top actor in the IT industry to offer his backing to Donald Trump. His transhumanist hopes, for instance for individual immortality, have been highlighted in articles about him in major US newspapers. He is also one of a small number of people in the IT industry who have given or still are giving material support to transhumanist activists. Other figures who have provided such support are the Russian Dmitry Itskov,[9] who is also interested in cosmism, and the Briton James Martin, who died in 2013 and donated large amounts of money to the University of Oxford, making it possible for philosophers with transhumanist leanings, like Nick Bostrom, to take up prominent posts there. The IT entrepreneur Larry Ellison (Oracle) has hopes of personally attaining immortality and funds research on biological ageing processes.

These examples are to be looked at against the background of the growth of a milieu with transhumanist sentiments since the 1970s, above all in the USA (Regis 1990; Schummer 2009; Coenen 2011; McCray 2012). Among the key figures in this were the controversial AI pioneer Mariv Minsky; Timothy Leary, who is best known as an "LSD guru" but was interested early on in cybernetic visions with wide-ranging implications and what is known as "cryonics"; the robotics expert Hans Moravec; and K. Eric Drexler, a futurist who had studied with Minsky and went on to popularize the term "nanotechnology". With the exception of Leary, these men can be termed (after Joachim Schummer) "visionary engineers" and (after Patrick McCray) "visioneers". With their sometimes audacious notions about the future, they have strongly influenced the image of R&D fields such as AI, nanotechnology and space research and, like Minsky's student Kurzweil, have written popular science books that depict very far-reaching visions of a posthuman future.[10] This milieu, which has been dubbed "The Third Culture" by the US publisher John Brockman, was introduced to a wider public in Germany thanks to the journalist Frank Schirrmacher, who died in 2014. In the early 2000s, Schirrmacher gave representatives of The Third Culture a platform in the review section of the influential German newspaper *Frankfurter Allgemeine Zeitung*.[11]

The transhumanist or pro-transhumanist milieu can be grasped as part of the dominant culture of digital capitalism, at the heart of which is found a mixture of technological optimism that often comes across as quasi-religious and a number of

[9]See the website of the 2045 Strategic Social Initiative: 2045.com.

[10]Many other scientists and engineers have helped to publicize such ideas, including Kevin Warwick, who is known more in the public sphere for his cyborg self-experiments and self-presentation than for his work on AI and robotics, and the academically very productive biochemist and molecular biologist George Church, who has gained greater prominence in the 2010s with transhumanist visions that go beyond biology.

[11]At that time, Schirrmacher cooperated with Brockman, the founder of the Edge network, whose collaborators also include the German publisher Hubert Burda. Brockman ran various events and publications that brought together well-known scientists with major players in the IT industry and other public figures. These activities are intended to promote the "The Third Culture", a network of intellectuals inspired by science to pursue innovative concepts. This network is closely linked with decision-makers in business and politics and has been heavily influenced by transhumanism.

central elements of the 1960s and 1970s "counterculture". This counterculture was by no means entirely hostile to or sceptical about technology and science but, with its prizing of creativity, its desire to break with the traditional, and its willingness to transgress boundaries, has had a crucial impact on IT culture. In recent times, these connections have received greater attention thanks to the personality cult around Steve Jobs (Apple). Jobs stated several times that the counterculture had been decisive for him, both personally and as far as his business philosophy was concerned. Furthermore, he once remarked that Gates suffered from the disadvantage of not having been influenced as much as he had himself by the consumption of LSD and other counterculture practices, saying he felt this meant Gates was "*unimaginative*".

4 Military Research and Transhumanism

What role did military research play in the development of transhumanism into a core ideological element of the culture of digital capitalism?

As discussed above, Bernal, probably the most important intellectual pioneer of transhumanism, and Good, who also influenced its evolution crucially with his concept of the intelligence explosion, made important contributions to the scientific war effort against National Socialist Germany. The same is true of the British-born US physicist Freeman Dyson, who was one of the most intellectually outstanding technofuturists of our time and has done a great deal to keep alive the memory of Bernal's significance as a technovisionary.

Yet, the thesis of an elective affinity between the MIC and transhumanism cannot be founded merely on the fact that a number of scientists with transhumanist visions of the future played valuable parts in work that made war more scientific and resulted in its computerization. What are more important here are the rise of AI, which capitalized on advances in cybernetics, and was pursued in the context of US military and space research during the Cold War, and the development of the cyborg subjectivity (blurring of the boundaries between human and machine) that is a core transhumanist vision (Edwards 1996) and is currently having a crucial impact on the civilian sector as well.

The elective affinity becomes very clear in the 1960s. After initial disappointment about the discrepancy between visions and results in the AI field, artificial intelligence was funded more generously again as a consequence of the shock caused by the Soviet Sputnik mission. From 1963, for example, an AI project at Massachusetts Institute of Technology (MIT) set up by, among others, the transhumanist thinker and AI pioneer Minsky received tens of millions of dollars for years from the military research budget (Grudin 2012, p. xxxvii). It is widely known that the history of the personal computer (PC) and the Internet goes back to US military research, in which respect Joseph Carl Robnett Licklider, the head of the Information Processing Techniques Office at the US Department of Defense's Advanced Research Projects Agency from 1962 to 1964, had a vital role as a science manager, while the engineer

Douglas Carl Engelbart paved the way for the digital world we live in today with the work he did towards the end of the 1960s at his Augmentation Research Center (Stanford University).

It has been mentioned above that the digital revolution on the US West Coast was influenced to a high degree by the contemporaneous countercultural milieu (Markoff 2005). Politically, however, some of the new digital elite distanced themselves sharply from the elements of the counterculture that were pacifist and critical of capitalism (Barbrook and Cameron 1996) and engaged in R&D activities that served the efforts being undertaken by the US MIC during the Cold War. More than anything else, however, this new elite was interested in making computers and the networks that interconnected them into tools for individual self-realization.

How much this counterculture-influenced technofuturism was shaped by the older transhumanism Bernal had espoused (which had been popularized in science fiction) is illustrated by the story of the L5 Society that was founded in 1975 (Regis 1990; Schummer 2009; Coenen 2011; McCray 2012). This organization of enthusiasts for the colonization of space drew, firstly, on Bernal's idea for a permanent space station ("Bernal Sphere"), which he had set out in his significant transhumanist essay *The World, the Flesh and the Devil* (1929). Secondly, the L5 Society—which was politically successful in its activism against the international Moon Treaty—provided a breeding ground for the development of modern transhumanism. Mention should be made here of the involvement of the nanotechnology futurist K. Eric Drexler, who subscribed to transhumanist views, and the transhumanist Martine Rothblatt (at that time still Martin Rothblatt) in the L5 Society. Furthermore, its development stands in an exemplary fashion for the partly ideological rapprochement of a new technoscientific elite (that was still heavily indebted to the "anti-establishment" elements of the counterculture) with the MIC. The question of the militarization of space became a central point of dispute within the L5 Society, in whose publications the countercultural enthusiasm for utopias of an alternative life away from the Earth was increasingly being displaced by an interest in military–industrial visions of space (McCray 2012, p. 140ff.).

The work of Stewart Brand is also indicative of the entwinement of the digital and psychedelic revolutions (see Turner 2006). In 1968, he assisted Douglas Carl Engelbart during a legendary presentation about the results of pioneering work on PC and Internet development that had been financed by US military research institutions. At the same time, he was a key figure in the psychedelic counterculture and published the magazine *Whole Earth Catalog*, which combined a do-it-yourself approach to technology with ecological and alternative-culture ideas and was later lauded by Jobs and other figureheads of digital capitalism.

Against the background that has been sketched out, it appears logical that one core term in contemporary transhumanism, "human (performance) enhancement", also has its roots in the IT industry, military research and the counterculture. Two aspects are of particular interest here: firstly, the ideas and technical developments in the field of human–computer interaction and AI that have been discussed above and were generated in military research—as well as the concept of the "cyborg" that came out of space research—prepared the ground for visions of a performance-oriented fusion of human and machine. Secondly, the concept of enhancing "human performance" that was central for the NBIC initiative was crucially popularized in a field where there were overlaps between military research and the counterculture.

As far as the first point is concerned, it is to be noted that not only were the technical foundations for the Internet and the PC laid by US military research into human–computer interaction and AI from the 1960s on but, for example, Licklider put forward a model of human–machine symbiosis, and Engelbart developed ideas about the coevolution of humans and computers that were acknowledged very prominently at events of the NBIC initiative. In Engelbart's approach, especially, it is a central proposition that human beings are able to augment their own intelligence by means of "communication" with advanced computer technology. In turn, this augmentation of intelligence will allow even better machines to be built, which will bring about yet further learning effects, and so on.

Furthermore, the concept of "human performance enhancement" has come to be of significance in other areas of military research since the mid-1980s. Potential ways of improving servicewomen and men's performance were investigated systematically under this label from 1984, with ideas and psychological techniques being appraised that had been developed in the countercultural "human potentials" movement. A Committee on Techniques for the Enhancement of Human Performance worked for 12 years on these largely psychological aspects of HE, examining scientifically questionable ideas from the counterculture as it did so (Druckman 2004).

The early 1990s then saw the biotechnological, neurotechnological and information technology "enhancement of human performance" being given greater weight in US military research. For instance, one report on what were at that time new scientific and technological developments argued that, since the soldier was a biological system, biotechnology offered unique potential for the enhancement of their performance (US National Research Council 1992, p. 151). The report predicted an extension of human performance by means of the direct coupling of the central nervous system to machines, as well as other "bionic" and orthopaedic developments. It was stated that "bionically" linked human–machine systems would be available around 2030. Further opportunities for HE were seen in the strengthening of the immune system with biotechnological and other methods. In the context of activities on nano-bio-info convergence, the "*new concept of enhancement (improving human performance)*" then found currency as a model in the second half of the 1990s (Smith 1998). The words in brackets subsequently came to be central for the NBIC initiative.

Unsurprisingly against the background that has been sketched out, the NBIC initiative itself, which was partly supported by US military research institutions, was characterized by a mingling of grandiose technological visions for the future of humanity with US military and economic interests, a combination that struck many as peculiar (particularly in Europe). At the same time, the initiative represented a political accolade for transhumanism (TAB 2008), even before its function as an ersatz religion for leading figures in the IT industry became widely apparent.

5 Military "Human Enhancement"?

This historical survey suggests it would be justifiable to speak if not of the birth of transhumanism from the spirit of military research (which itself has links with other traditions), but of the birth of the idea of "human (performance) enhancement" from that spirit.

For its part, military research has played a highly significant role in the more recent ethico-political debate about HE that began at the end of the 1990s, even though—very much in the tradition that has been outlined—many promoters of HE hold generally left-liberal and left-libertarian views, and therefore often ignore or play down the relevance of its military aspects. By contrast, a number of somewhat simplistic critiques have been put forward that depict the situation in black-and-white terms and suggest it is a mistake to see transhumanism's influence over leading IT industry figures and military research's openness to this worldview as unconnected aspects of the matter. For example, Jansen argues that both information technology, which he understands largely as military technology, and transhumanism are "*in essence*" concerned with the "*technological mastery and control of death*", in which respect "*only other people are ever supposed to die*" (Jansen 2015, p. 255).

This assessment is not only unfair, because it ignores the positive motivation behind modern transhumanism's desire to overcome death, but also misleading with regard to the contributions made by military research to the debate about HE. Generally, it can be concluded from the available literature as well as discussions involving experts in military research the author has been able to attend that—wholly in contrast to the "hype"-laden debate about ethical HE—a realistic view of the technical and pharmaceutical possibilities prevails in this field (cf. JASON 2008). This may be due above all to image considerations given the likelihood of problems gaining public acceptance for the "enhanced soldier" but does suggest that the specifically transhumanist visions of a human–machine symbiosis are regarded with some scepticism in military research. If at all, potential for effective HE is seen in the fields of pharmaceuticals (primarily to counter the decline in performance induced by lack of sleep) and nutrition.

Furthermore, examples can be found of efforts being made to reflect on the further-reaching implications of HE for military purposes. The present article will conclude by commenting on one attempt to do this, although it has to be allowed for that this is a publication by a retired serviceman, US Air Force Colonel Dave Shunk (2015).

Shunk begins his article by announcing that the "super soldier is on the way, maybe not tomorrow, but soon". Although this statement may be debatable, it has to be acknowledged that, when making it, he is concerned above all to emphasize the relevance of ethical considerations. Shunk starts by excluding technological developments that do not entail (surgical or pharmaceutical) interventions in the human body from the scope of his comments. However, he stresses that it will be possible for the soldier of the future to be "enhanced" by developments in neuroscience, biotechnology, nanotechnology, genetics and drugs. Here too, scepticism is called for, but this does not mean his reflections on ethical problems are irrelevant. Citing ancient myths of super soldiers and real experiments with drugs, those carried out by the German Wehrmacht during the Second World War for example, Shunk argues that the deployment of risky HE methods has long been an option available to the military. These comments are followed by a long series of questions about the challenges posed by the vision of HE. Pointing to the problematic use of other cutting-edge technology (drone attacks and the spying carried out by the US National Security Agency, NSA), he pleads for the ethical and legal aspects of HE to be discussed at an early stage. He ends his essay with Nietzsche's aphorism that, "He who fights with monsters should be careful lest he thereby become a monster." Since the 1980s, discussions of "human performance enhancement" in US military research have tended to focus on the danger posed by the possibility of enemy forces deploying super soldiers,[12] but this concern is at least questionable. After all, initiatives have come again and again from the US military research establishment itself to encourage positive engagement with the topic of HE. Nonetheless, the warning expressed with the quotation from Nietzsche definitely deserves to be taken on board. There is indeed great potential for transhumanist dreams to become true nightmares in the hands of the military.

The transhumanist worldview, in particular in its individualist variant influenced by the counterculture and ideas about the "space age", is not an indubitably dangerous or evil ideology, as fascism is for example. However, the fascination that may be exerted by wild dreams about making old human hopes come true must not lead to certain boundaries that need to be defended being lost sight of: just as it should be forbidden to use invasive HE methods (including drugs) on minors where this is not absolutely necessary (the child's right to physical integrity takes precedence over parents' rights here), non-therapeutic HE should also be banned in the military sector. In view of the structures of command and subordination that pervade the armed forces, there would otherwise be a danger of the right to physical integrity being undermined, with unforeseeable consequences for civilian life.

References

Altmann, J.: Nanotechnik, Singularität und Transhumanismus—Herausforderungen für Wissenschaft und Moral. FIfF Kommunikation 4, 50–52. https://www.fiff.de/publikationen/fiff-kommunikation/fk-2009/fk-4-2009/fiko_4_2009_altmann.pdf (2009)

Barbrook, R., Cameron, A.: The Californian ideology. Sci. Cult. **6**(1), 44–72 (1996)

Bernal, J.D.: The World, The Flesh and The Devil. Jonathan Cape, London (1970) (orig. 1929)

Clarke, A.C.: Glide Path. Harcourt, Brace & World, New York (1963)

[12]See Committee on Assessing Foreign Technology Development in Human Performance Modification (2012).

Coenen, C.: Verbesserung des Menschen durch konvergierende Technologien—Christliche und posthumanistische Stimmen in einer aktuellen Technikdebatte. In: Böhm, H., Ott, K. (eds.) Bioethik—Menschliche Identität in Grenzbereichen, pp. 41–124. Evangelische Verlagsanstalt, Leipzig (2008)

Coenen, C.: Extreme Technikvisionen als Verantwortungsproblem. In: Bartosch, U., Litfin, G., Braun, R., Neuneck, G. (eds.) Verantwortung von Wissenschaft und Forschung in einer globalisierten Welt, pp. 231–255. LIT, Berlin (2011)

Coenen, C.: Nachdarwinsche Visionen einer technischen Transformation der Menschheit. In: Ebert, U., Riha, O., Zerling, L. (eds.) Der Mensch der Zukunft, vol. 82/3, pp. 9–36. Hirzel (Abhandlungen der Sächsischen Akademie der Wissenschaften, Stuttgart (2013)

Coenen, C.: Transhumanism and its genesis. Humana. Mente. **26**, 35–58 (2014a). http://www.humanamente.eu/index.php/HM/article/view/114/96

Coenen, C.: Transhumanism in emerging technoscience as a challenge for the humanities and technology assessment. Teor. Praksa. **51**(5), 754–771 (2014b)

Committee on Assessing Foreign Technology Development in Human Performance Modification: Human Performance Modification—Review of Worldwide Research with a View to the Future. National Academy of Sciences, Washington, DC (2012)

Cross, T.: The novelist who inspired Elon Musk. 1843 Magazine, 31 Mar 2017. https://www.1843magazine.com/culture/the-daily/the-novelist-who-inspired-elon-musk (2017)

Druckman, D.: Be all that you can be—enhancing human performance. J. Appl. Soc. Psychol. **34** (11), 2234–2260 (2004)

Edwards, P.N.: The Closed World. MIT Press, Cambridge, MA (1996)

Ferrari, A., Coenen, C., Grunwald, A.: Visions and ethics in current discourse on human enhancement. NanoEthics. **6**(3), 215–229 (2012)

Good, I.J.: Speculations concerning the first ultraintelligent machine. In: Alt, F.J., Rubinoff, M. (eds.) Advances in Computers, vol. 6, pp. 31–88. Academic Press, New York (1965)

Groys, B., Hagemeister, M. (eds.): Die Neue Menschheit. Suhrkamp, Frankfurt (2005)

Grudin, J.: A moving target: the evolution of human-computer interaction. In: Jacko, J.A. (ed.) Human-Computer Interaction Handbook, 3. Aufl., S. xvii–lxi. CRC Press, Boca Raton (2012)

Jansen, M.: Digitale Herrschaft. Schmetterling, Stuttgart (2015)

JASON Defense Advisory Panel: Human Performance (JSR-07-625). JASON (The MITRE Corporation), McLean (2008)

Kurzweil, R.: The Singularity Is Near. Viking Penguin, New York (2005)

Levy, S.: The Plex—How Google Thinks, Works, and Shapes Our Lives. Simon & Schuster, New York (2011)

Markoff, J.: What the Dormouse Said—How the Sixties Counterculture Shaped the Personal Computer Industry. Viking, New York (2005)

McCracken, H., Grossman, L.: Google vs. Death. Time, 30 Sept 2013 (title story) (2013)

McCray, W.P.: The Visioneers. Princeton University Press, Princeton (2012)

McCray, W.P.: Futures perfect and visioneering—a re-assessment. NanoEthics. **11**(2), 203–207 (2017)

Millikan, A.: I am a cyborg and I want my google implant already. The Atlantic, 30 Sept 2010. https://www.theatlantic.com/technology/archive/2010/09/i-am-a-cyborg-and-i-want-my-google-implant-already/63806/ (2010)

Nordmann, A.: If and then—a critique of speculative nanoethics. NanoEthics. **1**(1), 31–46 (2007)

Parrinder, P.: Shadows of the Future. Liverpool University Press, Liverpool (1995)

Pièces et Main d'Oeuvre: Transhumanisme—Du progrès de l'inhumanité. Online-Dokument. http://www.piecesetmaindoeuvre.com/IMG/pdf/Transhumanisme_inhumanite_-2.pdf (2015)

Regis, E.: Great Mambo Chicken and the Transhuman Condition. Perseus, Reading (1990)

Schneider, S.: Future minds—transhumanism, cognitive enhancement, and the nature of persons. In: Ravitsky, V., Fiester, A., Caplan, A.L. (eds.) The Penn Center Guide to Bioethics, pp. 95–100. Springer, New York (2009)

Schummer, J.: Nanotechnologie—Spiele mit Grenzen. Suhrkamp, Frankfurt (2009)

Shunk, D.: Ethics and the enhanced soldier of the near future. Mil. Rev. **95**, 91–98 (2015)
Slusser, G.: Dimorphs and doubles—J.D. Bernals "two cultures" and the transhuman promise. In: Westfahl, G., Slusser, G. (eds.) Science Fiction and the Two Cultures, pp. 96–130. McFarland, Jefferson (2009)
Smith II, R.H.: A policy framework for developing a national nanotechnology program. Thesis, Virginia Polytechnic Institute and State University, S.50. http://scholar.lib.vt.edu/theses/available/etd-101698-094822/unrestricted/thesis.pdf (1998)
TAB, Büro für Technikfolgen-Abschätzung beim Deutschen Bundestag: Konvergierende Technologien und Wissenschaften (Author: Coenen, C.). TAB, Berlin. https://www.tab-beim-bundestag.de/de/pdf/publikationen/berichte/TAB-Hintergrundpapier-hp016.pdf (2008)
Turner, F.: From Counterculture to Cyberculture. University of Chicago Press, Chicago (2006)
US National Research Council, Board on Army Science and Technology: STAR 21—Strategic Technologies for the Army of the Twenty-First Century. National Academies of Sciences, Washington, DC (1992)
Vogeler, R.-D.: Engagierte Wissenschaftler. Peter Lang, Frankfurt (1992)
Wagner, T.: Robokratie. PapyRossa, Köln (2015)
Werskey, G.: The visible college revisited—second opinions on the red scientists of the 1930s. Minerva. **45**(3), 305–331 (2007)

Chapter 7
When CRISPR Meets Fantasy: Transhumanism and the Military in the Age of Gene Editing

Robert Ranisch

Abstract Newly discovered tools for gene editing such as CRISPR allow direct modification of the DNA of organisms. This could not only make new therapeutic applications possible. In theory, gene editing could also be used to enhance human beings or even to modify the human germline, i.e. inducing changes that could be inherited by future generations. Considering these possibilities, it comes as no surprise that the discovery of CRISPR was greeted with euphoria in transhumanist circles. Germline interventions are seen as a possible key for a posthuman future. As it will be argued here, this popular perception is based on an overly simplistic understanding of genetics and exaggerated expectations of the potential of gene editing technologies. In addition, fantasies about human enhancement also resemble emerging speculations about "super soldiers" for future warfare. While it will be maintained that genetically upgraded combatants ought to be relegated to the realm of science fiction, applications of gene editing technologies in the military should still cause concerns about biosecurity risks.

1 Introduction

Transhumanism in a narrow sense can be described as an ideology and a movement[1] that aims to apply new technologies to change human characteristics so radically that we can talk of an evolutionary jump forward (Ranisch and Sorgner 2014, 7–9). The

[1]For instance, Humanity+ (h+) as well as the Institute for Ethics and Emerging Technologies (IEET) are counted among the most well-known transhumanist organizations. On the transhumanist movement, see Ranisch and Sorgner (2014, 12–13).

R. Ranisch (✉)
Research Unit "Ethics of Genome Editing", Institute for Ethics and History of Medicine, University of Tübingen, Tübingen, Germany

International Centre for Ethics in the Sciences and Humanities (IZEW), University of Tübingen, Tübingen, Germany
e-mail: robert.ranisch@uni-tuebingen.de

W. Hofkirchner, H.-J. Kreowski (eds.), *Transhumanism: The Proper Guide to a Posthuman Condition or a Dangerous Idea?*, Cognitive Technologies,
https://doi.org/10.1007/978-3-030-56546-6_7

transhumanist motif delineates the transition of the human to the posthuman. Beyond eschatological ideas to take human evolution into a posthuman age, we can also talk of transhumanist technologies or developments in a broader sense. These encompass means which promise to improve the physical and mental capabilities as well as the behaviours of humans by working directly on or in the body (*human enhancement*). In this regard, applications from the areas of nano-, bio-, and information technologies as well as cognitive science (NBIC) are especially noteworthy.

Due to the anticipated potential to increase human capacity, a strong interest of the military in the research and development of respective technologies is hardly surprising (*military human enhancement*). While self-proclaimed transhumanists rarely present themselves as avid supporters of the further development of military technology (cf. Švaňa 2017), it does carry a special value for them: More than in the civilian sphere, here it is visible how systematic efforts and well-financed research can generate technologies for improvement beyond normal human possibilities. For these reasons, military research and development is associated with a high transhumanist potential (Thomas 2017).

2 Gene Editing: The "CRISPR Revolution"

One of the newest options acclaimed as a potential transhumanist technology may be methods of gene editing that have just recently been developed. Particularly, the CRISPR technology not only spurs the imagination of natural scientists (for basic introduction: Doudna and Charpentier 2014). It also finds great affirmation in the current discussions of transhumanists and techno-progressive thinkers (e.g. de Araujo 2017; Sorgner 2018; Hughes 2015).

Although technologies for directly changing the DNA of organisms have been in use for several decades, their application has remained fairly limited. Likewise, the persistent hopes for medical developments of gene therapies, such as against cancer or autoimmune diseases, have not come to fruition yet. Despite comprehensive endowments, financial and otherwise, in this area, from three decades of research only a fraction of the studies led to applications fit for clinical use. This could change with the new techniques of gene editing. CRISPR is seen as a disruptive technology which supplants competing methods from labs within the shortest time. Common opinion holds that these new possibilities of gene editing are significantly cheaper, more effective and easier to use than previous methods.

Since CRISPR can be applied to any type of cell, the potential applications of gene editing are manifold. Within the shortest time this new technology was used on plants, fungi and animals (see, e.g., Nuffield Council on Bioethics 2016). The development of new fuels, materials or pharmaceutical products is also in the scope. Using CRISPR it could be possible to alter animals and make their organs suitable for humans. In connection with the so-called *gene-drive* method, interventions in entire ecosystems are conceivable. For instance, research is being done on changing the anopheles mosquito which would lead to an extermination of the population within a few generations and so be omitted as a carrier of malaria. If only a few such organisms, who were changed, left the laboratory, it could have a large-scale impact.

The greatest hopes and concomitantly the greatest fears relate to the application of gene editing on humans or human cells. Two types of gene editing must be distinguished: somatic and germline interventions. In the first case, somatic cells (i.e. cells of the human body other than sperm or egg cells) are targeted, e.g. to treat diseases of a patient such as beta thalassemia. During the last few years, we witness a competition between the USA and China for the first clinical applications of CRISPR gene therapies. In the second case, germline gene editing targets human germ cells or early embryos. Should altered embryos be implanted and carried to full term, respective changes would be inherited by future generations.

Germline interventions are highly controversial. Alternating human embryos for reproductive purposes is widely disdained for safety reasons but also because of the fear of new forms of "eugenics" (Ranisch 2019), resulting in social inequalities, and possible dual use (van Dijke et al. 2018). After Chinese scientist altered (non-viable) human embryos in vitro for the first time in 2015 (Liang et al. 2015), an international moratorium for this kind of research was widely discussed (e.g. Lanphier et al. 2015). Only a few months later, however, US-American and European research teams followed with similar experiments on in vitro embryos. If there is any kind of identifiable consensus, then it would be too early to use germline gene editing for reproductive purposes and that germline intervention must not be used for non-medical purposes.

In November 2018, however, Chinese scientist He Jiankui revealed that he created genetically modified humans (Dzau et al. 2018). An unauthorized experiment resulted in the birth of two girls, one of them supposedly carries a genetic mutation, making her resistant to human immunodeficiency virus (HIV) infection. If this story proves to be true, this would be the first instance of a targeted germline intervention in humans using the CRISPR technology. Most remarkably, He Jiankui's widely condemned experiment had no therapeutic but a preventive purpose: he aimed at enhancing immunity of future offspring.

3 Transhumanist Visions: Enhancing Human Evolution

In transhumanist circles, the discovery of new gene editing technologies was greeted with euphoria. Some even consider CRISPR to be the "most powerful technological invention of this decade" (Sorgner 2018). Especially germline intervention using CRISPR technique is widely held to be the key for posthuman change (cf. Porter 2017). In the dawn of the first germline interventions, leading transhumanist James Hughes confirmed the official 2004 statement from the World Transhumanist Association which highlights the "desirability and inevitability of germline and enhancing gene therapies" (Hughes 2015).

The reason why germline gene editing is associated with a high transhumanist potential is the lack of existing technologies for successful genetic enhancement. Pre-implantation genetic diagnostics (PGD) and embryo selection, which is being used for more than 25 years, allow selection of embryos with desired genetic traits. But this technology is mainly effective in avoiding genetic disease which is caused by a mutation on a single gene (monogenic). While PGD can help to identify and select out embryo that carries such a deleterious mutation, it is highly unlikely that genetic screening could detect some kind of "transhuman" features in early human life (Ranisch 2020).

Almost all complex human traits on the transhumanist's wish list are caused by numerous factors. When authors speculate to augment intelligence, cognitive abilities or bodily strength, no single variant of a gene would have any significant effect. Hence, established technologies such as PGD would be of little help. Although embryo selection for polygenic traits is technically feasible, it would only have modest enhancing effects. Even *if* all the genetic factors e.g. for complex traits such as intelligence were known, a massive number of embryos were necessary to enhance future offspring (Shulman and Bostrom 2014).

Now, with the advent of germline gene editing, transhumanists and techno-progressive thinkers consider polygenic gene editing to be a realistic scenario. Christopher Gyngell and his colleagues speculate that targeting multiple genes in human embryos could significantly reduce the risk of diabetes, coronary artery disease or cancer by targeting multiple genes, making germline gene editing a "powerful disease-preventing technology" (Gyngell et al. 2017, 501), which could radically extend life- and health span. Marcelo de Araujo (2017) goes one step further. He argues that germline gene editing for cognitive enhancement may be necessary in the future and that "policy makers may even come to consider genetic cognitive enhancement as a matter of public policy". Transhumanist Zoltan Istvan, who ran for US president in 2016, argues in a similar vein. Since he has no doubts that China will use gene editing to radically enhance its citizens, the USA should "just embrace genetic editing and be better at it than the Chinese" (Istvan 2016). Ironically, he sees this genetic arms race as a guarantee for peace. If only the Chinese are enhanced, the world may end up in a new cold war between the "old-fashioned humans" and the "supermans".

4 Gene Editing in a Military Context

Considering the possible therapeutic and imagined enhancement effect on human application of gene editing techniques, it inevitably caught the attention of the military. Especially in connection with innovations regarding military medicine, a number of military uses can be conceived:

> Areas of potential interest include research aimed at improving battlefield medicine and the acceleration of basic research into physiological and psychological responses to trauma, healing mechanisms and the development of related products and treatments. More

> speculatively, there is also potential interest in employing genome editing for the enhancement of personnel, in relation to genetic susceptibilities to conditions that they might experience in warfare, improving concentration, and other physiological characteristics such as physical fitness. The most evident security interest, however, is in identifying and countering external threats. (Nuffield Council on Bioethics 2016, 111)

Several studies from the last few years addressed NBICs in the military context and how these could change warfare and which ethical, social and legal challenges are connected with that (Beard et al. 2016; Carrick et al. 2013; Harrison Dinniss and Kleffner 2016; Lin et al. 2013; Mehlman 2015). Parallels between new means of warfare and transhumanist visions are drawn not only by the news media but also by researchers (Švaňa 2017; Al-Rodhan 2015). These comparisons focus less on the circumstance that transhumanism might have a strong interest in military questions, but rather on the radicalness and degree of intervention with which recent military technology aims to increase the performance of and enhance the human body.

In this regard, the public media and participants in the academic discourse often speak of so-called *super soldiers* (e.g. Galliott and Lotz 2016). This quasi-transhuman figure, which appears in the literature as a superhuman being or as a mindless killing machine, is increasingly associated with the possibilities of germline gene editing. Even security experts are sounding a warning call. Some consider the possibility of editing human embryos as the first steps towards such genetically modified superhumans: "you could, at least in principle, make [. . .] a soldier with great muscle force and strength" (qtd. in: Ansley 2016). Foreign policy analyst Brent M. Eastwood recently claimed that "China is clearly pursuing dual-use genetic engineering technology" and stated that "CRISPR could someday enable U.S. adversaries to [. . .] create 'super soldiers' to dominate future battlefields" (Eastwood 2017).

The use of technologies for gene editing to enhance the human body must be viewed especially critically. This goes even more so for future visions of super soldiers being "bred" in secret laboratories—often projected onto China by the Western media (cf. Schaefer 2016). Even if we assumed that such research, which would result in a systematic breach of human rights considering the large-scale experimentation on humans, was not limited by any research ethics (Greene and Master 2018), there are grave doubts from a scientific and pragmatic view regarding the plausibility of such a dystopia.

Beyond simplistic deterministic understanding of genetics, it is highly unlikely that the genetic factors for relevant features could all be distinctly identified and respectively targeted for modification. While in recent years, genome-wide association study (GWA) made progress in identifying some genetic contributions to factors such as intelligence or even risk-taking behaviour, two insights can be gained from these studies: Firstly, genetic variants can only explain a fraction of relevant features. Secondly, GWA suggest that complex traits are associated with many genetic variants. Even supposedly simple phenotypical features such as body height are influenced by several thousand factors.

In particular this second insight shows the current limitations of germline gene editing to enhance complex traits. Despite rapid progress in this field, there are no

Table 7.1 Variants of genes associated with "transhuman qualities" (source: Church Lab 2017, selection)

Variant of gene	Protective, resilient or extreme effects
LRP5	Extra-strong bones
MSTN	Large, lean muscles, low atherosclerosis
SCN9A	Insensitivity to pain
CCR5	HIV resistance
FUT2	Norovirus resistance
HBB	Malaria resistance
BDKRB2	Deep diving
PDE10A	Breath-hold diving
EGLN1	High altitude
EPOR	High oxygen transport
DEC2	Less sleep

indications that the necessary precision and efficacy for changing multi loci in the germline is possible soon. Basic research rather shows that the CRISPR method often does not function as precisely as hoped, even when only a single gene is targeted. The current state of research suggests that the therapeutic use of gene editing to modify the germline is in its infancy and still years away from clinical applications. Intervening in the germline for the purpose of radical enhancement, i.e. changing complex traits, is speculation at this point. Apprehensions that we need to beware of radically enhanced genetically modified descendants are premature.

If germline gene editing is used in the near future, then it is only to target monogenic traits. In fact, there are a few known naturally occurring variants of single genes associated with specific qualities including lean muscles or strong bones as well as resistance to some infectious diseases. Harvard geneticist George Church, sometimes considered to be the "most famous of transhumanist scientists" (Knoepfler 2016, 185), is eager to spot these variants (Table 7.1). This list could provide a blueprint for transhumanists or the military for possible germline interventions using CRISPR gene editing technique.

Even though Church sometimes suggests that these genetic variants count as "transhuman" qualities (Church and Regis 2014, 228), most of these variants either have only modest effects or carry costs: Mutations in the SCN9A gene, which are associated with an insensitivity to pain, are extremely dangerous for those affected, because injuries go unnoticed. High bone density could be beneficial for athletes but a disadvantage for swimmers. Even the variant of the CCR5 gene that can protect from HIV infections could have detrimental effects, since it makes humans vulnerable to infection by West Nile virus. Considering the high-flying ideas of genetically enhanced transhuman soldiers, these single gene variants are rather uninteresting.

Furthermore, even if the safety and efficacy of gene editing technologies changed in the future, pragmatic reasons speak against germline intervention for military purposes. Several decades would pass until such "super soldiers" could be sent into battle while the technologies employed for this purpose would likely be outdated by then (cf. Sparrow 2015). Should such genetically enhanced soldiers have a higher endurance and intelligence, they would still be humans made of flesh and blood; it is doubtful whether these could provide the envisioned advantages in warfare a few decades from now.

A more realistic scenario, however, would be interventions using gene editing technologies in living persons. It is an open secret that the military has shown a keen interest in various forms of pharmacological means to improve performance (Kamienski 2016). While drugs and narcotics have long been in use to (supposedly) enhance performance, there had been attempts to generate biotechnological enhancements in the past (Ford and Glymour 2014). A quick perusal of the freely available documentation by US-American Defense Advanced Research Projects Agency (DARPA) provides sufficient confirmation: There research projects to enlarge the capabilities of soldiers are mentioned such as making grass digestible for humans, handling unusual levels of strain and stress, communicating quasi-telepathically or attaining improved sensory perception (Lin 2016, 60–61). In this context, CRISPR could be one new tool in the box of military medicine. For example, future developments could protect soldiers against biological agents (Greene and Master 2018), and further military research geared towards an increase in physical and mental capacities. Like cases of gene doping, endurance or muscle growth could be promoted. In animal experiments, it was already possible to increase the muscle mass of dogs using CRISPR.

Even for the use of gene editing on living soldiers, we should be sceptical whether any significant enhancing effects could be expected: long-term improvements normally do not exist without risks or without being constrained by the "natural" limits of the human body (cf. Bostrom and Sandberg 2009). Furthermore, considering the average time frame of more than two decades until basic research leads to clinical applications, it needs to be asked whether improvements of the human body or psyche could not also be attained with less risky, non-invasive and reversible technologies.

5 Outlook: Focusing on Biohazard Threats and Not Super Soldiers

This sceptical stance towards the idea of future warfare being conducted by genetically upgraded super soldiers does not mean that gene editing technology does not pose future safety risks. The relatively easy and low-cost application of CRISPR and the harmful potential of gene editing inevitably caught the attention of the military and of experts of defence. For instance, gene editing was discussed by the national director of the US Secret Service, James R. Clapper, as a potential weapon of mass destruction, which could be a threat to national security if it fell into the wrong hands (Clapper 2016, 9). This seems to address the dangers which result from the possible development of new biological weapons and the abuse of the *gene-drive* method. Do-it-yourself CRISPR sets available from the Internet already allow lay scientists or biohackers to perform easy experiments at the very least. Considering the low threshold for using these new technologies, from the perspective of biosecurity and hazard control, these new technologies pose very particular challenges and risk profiles (Fears and ter Meulen 2018).

In turn, the DARPA has set up the project *Safe Genes* "to build a biosafety and biosecurity toolkit to reduce potential risks and encourage innovation in the field of genome editing" (DARPA 2016). DARPA also supports research on gene editing at several leading US-American universities (DARPA 2017). It must be assumed that the military interest in gene editing is not limited to counter- or defence measures (Reeves et al. 2018), although a detailed analysis of current military projects naturally is hardly possible.

There are justified concerns regarding possible bioterrorism in connection with the gene editing and gene-drive method (Ahteensuu 2017; Gurwitz 2014). But genetically modified, transhuman soldiers ought to be relegated to the realm of science fiction. A factual parallel between transhumanist ideas and current discussions about super soldiers is in evidence though: Both discourses are preoccupied with speculative, sometimes dire future scenarios that reveal a utopian or dystopian way of thinking that is removed from scientific analysis. Admittedly, caution may be recommended considering the speed of past developments, especially of military technology. History has shown that many existing innovations did not seem feasible only a short time before. However, to come to the reverse conclusion could be just as misguided: not everything imaginable becomes a reality.

References

Ahteensuu, M.: Synthetic biology, genome editing, and the risk of bioterrorism. Sci. Eng. Ethics. (2017). https://doi.org/10.1007/s11948-016-9868-9

Al-Rodhan, N.: Transhumanism and war. In: Global Policy. www.globalpolicyjournal.com/blog/18/05/2015/transhumanism-and-war (2015)

Ansley, R.: Gene editing: the good, the bad, and the ugly. In: Atlantic Council. www.atlanticcouncil.org/blogs/new-atlanticist/gene-editing-the-good-the-bad-and-the-ugly (2016)

Beard, M., Galliott, J., Lynch, S.: Soldier enhancement: ethical risks and opportunities. Aust. Army J. **13**(1), 5–20 (2016)

Bostrom, N., Sandberg, A.: The wisdom of nature: an evolutionary heuristic for human enhancement. In: Savulescu, J., Bostrom, N. (eds.) Human Enhancement, pp. 375–416. Oxford University Press, Oxford (2009)

Carrick, M.D., Connelly, J., Lucas, G., Robinson, P.: New Wars and New Soldiers: Military Ethics in the Contemporary World. Ashgate Publishing, Burlington, VT (2013)

Church Lab. Multigenic traits can have single genes with large impacts. http://arep.med.harvard.edu/gmc/protect.html (2017)

Church, G.M., Regis, E.: Regenesis: How Synthetic Biology Will Reinvent Nature and Ourselves. Basic Books, New York (2014)

Clapper, J. R.: Statement for the record: worldwide threat assessment of the US Intelligence Community. Senate Armed Services Committee. https://www.dni.gov/files/documents/SASC_Unclassified_2016_ATA_SFR_FINAL.pdf (2016)

DARPA. Setting a safe course for gene editing research. https://www.darpa.mil/news-events/2016-09-07 (2016)

DARPA. Building the safe genes toolkit. https://www.darpa.mil/news-events/2017-07-19 (2017)
de Araujo, M.: Editing the genome of human beings: CRISPR-Cas9 and the ethics of genetic enhancement. J. Evol. Technol. **27**(1), 24–42 (2017)
Doudna, J.A., Charpentier, E.: The new frontier of genome engineering with CRISPR-Cas9. Science. **346**(6213), 1258096 (2014)
Dzau, V.J., McNutt, M., Bai, C.: Wake-up call from Hong Kong. Science. **362**(6420), 1215 (2018). https://doi.org/10.1126/science.aaw3127
Eastwood, B. M.: Gene-editing in China: beneficial science or emerging military threat? In: Atlantic Council. www.atlanticcouncil.org/blogs/futuresource/gene-editing-in-china-beneficial-science-or-emerging-military-threat (2017)
Fears, R., ter Meulen, V.: Assessing security implications of genome editing: emerging points. From an international workshop. Front. Bioeng. Biotechnol. **6** (2018). https://doi.org/10.3389/fbioe.2018.00034
Ford, K., Glymour, C.: The enhanced warfighter. Bull. Atomic Sci. **70**(1), 43–53 (2014)
Galliott, J., Lotz, M. (eds.): Super Soldiers: The Ethical, Legal and Social Implications. Routledge, New York (2016)
Greene, M., Master, Z.: Ethical issues of using CRISPR technologies for research on military enhancement. Bioeth. Inq. **15**, 327–335 (2018)
Gurwitz, D.: Gene drives raise dual-use concerns. Science. **345**(6200), 1010 (2014)
Gyngell, C., Douglas, T., Savulescu, J.: The ethics of Germline gene editing. J. Appl. Philos. **34**(4), 498–513 (2017)
Harrison Dinniss, H.A., Kleffner, J.K.: Soldier 2.0: military human enhancement and international law. Int. Law Stud. **92**(1), 432–482 (2016)
Hughes, J.: Transhumanist position on human germline genetic modification. In: IEET. https://ieet.org/index.php/IEET2/more/hughes20150320 (2015)
Istvan, Z. (2016) Genetic editing could cause the next cold war. https://www.vice.com/en/article/ezp8me/genetic-editing-could-cause-the-next-cold-war
Kamienski, L.: Shooting Up: A Short History of Drugs and War. Oxford University Press, Oxford (2016)
Knoepfler, P.: GMO Sapiens: the Life-Changing Science of Designer Babies. World Scientific, Hoboken, NJ (2016)
Lanphier, E., Urnov, F., Haecker, S.E., Werner, M., Smolenski, J.: Don't edit the human germ line. Nature. **519**(7544), 410–411 (2015). https://doi.org/10.1038/519410a
Liang, P., Xu, Y., Zhang, X., Ding, C., Huang, R., Zhang, Z., Huang, J.: CRISPR/Cas9-mediated gene editing in human tripronuclear zygotes. Protein Cell. **6**(5), 363–372 (2015). https://doi.org/10.1007/s13238-015-0153-5
Lin, P.: Ethical blowback from emerging technologies. In: Demy, T.J., Lucas Jr., G.R., Strawser, B.J. (eds.) Military Ethics and Emerging Technologies. Routledge, New York (2016)
Lin, P., Mehlman, M., Abney, K.: Enhanced warfighters: risk, ethics, and policy. http://ethics.calpoly.edu/Greenwall_report.pdf (2013)
Mehlman, M.J.: Captain America and Iron man: biological, genetic, and psychological enhancement and the warrior ethos. In: Lucas, G. (ed.) Routledge Handbook of Military Ethics, pp. 406–420. Routledge, New York (2015)
Nuffield Council on Bioethics. Genome editing: an ethical review. http://nuffieldbioethics.org/wp-content/uploads/Genome-editing-an-ethical-review.pdf (2016)
Porter, A. Bioethics and transhumanism. The journal of medicine and philosophy: a forum for bioethics and philosophy of medicine vol. 42, no. 3, pp. 237–260 (2017)
Ranisch, R.: 'Eugenics is back'? historic references in current discussions of germline gene editing. NanoEthics. **13**(3), 209–222 (2019)
Ranisch, R.: Germline genome editing versus preimplantation genetic diagnosis: Is there a case in favour of germline interventions? Bioethics. **34**(1), 60–69 (2020)
Ranisch, R., Sorgner, S.L.: Introducing post- and transhumanism. In: Ranisch, R., Sorgner, S.L. (eds.) Post-and Transhumanism: An Introduction, pp. 7–27. Peter Lang, Frankfurt (2014)

Reeves, R.G., Voeneky, S., Caetano-Anollés, D., Beck, F., Boëte, C.: Agricultural research, or a new bioweapon system? Science. **362**(6410), 35–37 (2018)
Schaefer, G. O.: China will develop the first genetically enhanced 'superhumans', experts predict. www.dailymail.co.uk/sciencetech/article-3721991/China-develop-genetically-enhanced-superhumans-experts-predict.html (2016)
Shulman, C., Bostrom, N.: Embryo selection for cognitive enhancement: curiosity or game-changer? Global Policy. **5**(1), 85–92 (2014)
Sorgner, S.L.: Genes, CRISPR/Cas 9, and posthumans. In: Ethics of Emerging Biotechnologies, pp. 5–17. Trivent Publishing, Budapest (2018)
Sparrow, R.: Enhancement and obsolescence: avoiding an "enhanced rat race". Kennedy Inst. Ethics J. **25**(3), 231–260 (2015)
Švaňa, L.: (Military) human enhancement–ethical aspects. Hum. Aff. **27**(2), 155–165 (2017)
Thomas, A.: Super-intelligence and eternal life: transhumanism's faithful follow it blindly into a future for the elite. In: The Conversation. https://theconversation.com/super-intelligence-and-eternal-life-transhumanisms-faithful-follow-it-blindly-into-a-future-for-the-elite-78538 (2017)
van Dijke, I., Bosch, L., Bredenoord, A.L., Cornel, M., Repping, S., Hendriks, S.: The ethics of clinical applications of germline genome modification: a systematic review of reasons. Hum. Reprod. (2018). https://doi.org/10.1093/humrep/dey257

Chapter 8
War in Times of "Beyond Man": Reflections on a "Grand" Contemporary Topic

Alexander Reymann and Roland Benedikter

Abstract This chapter argues that, although transhumanism depicts itself as following the footsteps of the tradition of humanism, i.e. as seeking for a better world for all human beings, it is realizing ambiguous trajectories. Despite officially proclaimed integrative intentions and goals of some of its most prominent leaders, it seems that transhumanism tends to push forward social innovations that are a double-edged sword. Indeed, we face an era of military rearmament also due to achievements that have emerged in the converging fields of AI, robotics and human enhancement. Some of the most controversial views concerning these achievements and their possible impacts on modern warfare are discussed. Moreover, we outline how these developments in the military sector are symptomatic for a broader technological trend that seems to become a major transformative driver in the twenty-first century—a world where inequality is on the rise both in the social, the technological and the military spheres.

1 Introduction

Technology is increasingly shaping individual and social life. Going back, if desirable at all, seems impossible—and technological and scientific progress seems unstoppable. Technology has become the third social force besides politics and economics—indeed, its timely significance begins to transcend and overtake that of the other two.

Based on such observations, "transhumanism" has formed as a loose philosophical and sociopolitical current. The programme of the transhumanist movement is to

A. Reymann
Görgeshausen, Germany

R. Benedikter (✉)
Center for Advanced Studies, Eurac Research, Bozen-Bolzano-Bulsan, Italy

Multidisciplinary Political Analysis, Willy Brandt Centre, University of Wroclaw, Wrocław, Poland

W. Hofkirchner, H.-J. Kreowski (eds.), *Transhumanism: The Proper Guide to a Posthuman Condition or a Dangerous Idea?*, Cognitive Technologies,
https://doi.org/10.1007/978-3-030-56546-6_8

bring the increasing importance of technology and the extent of social change it brings about in the focus of public perception and to put it on the worldwide political agenda more than ever before. For some years now, it has begun to organize itself politically as a loose, transnational party network, with foundations in the USA, Great Britain and starting in Germany as well. In 2016, the "U.S. Transhumanist Party" campaigned for US presidency. The candidate was the intellectual, journalist and entrepreneur Zoltan Istvan, author of the best-selling book *The Transhumanist Wager* (2013).

There is no denying the ever-increasing need to discuss what is already technically feasible in a broader societal perspective and to regulate it politically and legally. The almost exclusive focus of transhumanism on how to evaluate and manage the development at the point of intersection of man and technology and deriving the solution to all other issues from that intersection gives the movement its unique selling proposition. It is characterized by a fundamentally optimistic attitude towards a technology-permeated future: nothing less than the fundamental transformation and reorganization of human life and living together—robots that will soon take over painstaking work from people, intelligent homes and communicating and learning devices that organize our lives. Transhumanists also advocate self-driving cars, holidays in virtual reality or on the moon, and they are preparing for the colonization of space, including—according to Elon Musk, the co-founder of Space-X and the thinker and driver behind the trendy electric automotive brand Tesla—the colonization of Mars.

But beyond that, transhumanism is much more: It is about the change of man himself, his body, his spirit, his nature, his (self-) conception. From a transhumanist point of view, "change" always means improvement, i.e. human enhancement. The optimization of man and his "conditio humana" through the radical use of technology is the unifying leitmotif of the movement, which is otherwise diverse, with contentious left and right wings, and heterogeneous down to its foundations.

2 The Core Contradiction: Healing the Sick or "Enhancing" the Healthy Person?

Without question, it is a noble goal of transhumanist efforts to enable the blind to see and the deaf to hear in the foreseeable future, for example by means of retina chips or cochlear implants, i.e. an "intrusive" technology that fuses with the human body and, according to the transhumanists' vision, should make it increasingly interchangeable, so that in the ideal case more and more parts of it should be replaceable. Similarly, intelligent pacemakers, sensitive prostheses and other technical aids should alleviate the suffering of those affected. Improving the quality of life through comprehensive transdisciplinary and transversal scientific progress is an avowed part of the transhumanist programme.

But it is only starting from the condition of healthy humans that the transhumanist utopia takes on its true contours and unfolds its constitutive conflict potential, characteristic of the transhumanist ideology. Humans can be "better"; this is the credo—smarter, faster, more robust, more networked and, last but not least, maybe even immortal, if humans are upgraded and modified more or less unconditionally and radically with technology, and thereby transcend their previous self-exploration experience, entering into unknown areas. This in principle applies to healthy humans as much as to the sick ones, indeed even much more. For in principle, as the transhumanist maxim goes—often accompanied by polemics—a "higher level" of enhancement can be achieved starting from a healthy body.

In essence, transhumanism actually dreams of a "new human being", whom it conceives not as an individual but as a "new humanity" (neo-humanity), as the "Global Future Congress 2045" formulated in March 2013 in an open letter to the then UN Secretary-General Ban Ki-moon (Global Futures 2045). In that letter, world-leading philanthropists such as James Martin, co-founder and namesake of the James Martin 21st Century School at Oxford University, and scientists such as Anders Sandberg demanded a widespread redeployment of public funds into the technology sector, including for the creation of a man–technology continuum from which one should expect a whole new human condition and quality of being.

The ways and means to reach this goal are manifold. Some are already a reality—such as the targeted alteration of the genome or the direct brain–computer and brain–machine interfaces, in which man, artificial intelligence and machine connect into one single unit. Other projects are ambitious, but tangible, such as the head transplant, probably first to be attempted in 2017 (Benedikter et al. 2017). Still others, at least from today's perspective, are still a long way off, in particular concrete ways to overcome death, but also the lasting interruption of cellular ageing processes, the extensive cyborgization of the body or the decoupling of body and mind with subsequent digitization and implementation of the mind into a non-biological substrate, also known as "mind-uploading" (ibid). Logical and philosophical inconsistencies are seemingly ignored here (it is at least doubtful whether a separation of body and mind can be more than a Cartesian fantasy).

However, the seriousness of the transhumanists in achieving their goals as a true "leap in quality for humanity" is demonstrated also by the visions (frenzy-raising and widely discussed internationally for years now) of their prominent representatives, such as Nick Bostrom, Director of the "Future of Humanity" Institute at Oxford University and Advisor to the BRAIN Project of the US Government under Barack Obama, or Ray Kurzweil, Head of Technical Development at the nominally most valuable business in the World, Google. Already with these VIPs involved in the movement, one thing becomes clear: transhumanism is not a niche ideology by and for science fiction lovers. It is an ideology that is borne and spread by influential and financially strong actors from politics, business and research. It is no coincidence that a centre of the movement is located in the technology forge of the USA, the Silicon Valley. And one of their most important institutions to date and a vision that has come to life—the Singularity University, founded in 2008 by the entrepreneur and aeronautical engineer Peter Diamandis and by Ray Kurzweil, dedicated to preparing for the expected awakening of technology to self-consciousness (singularity)—is located on the campus of the former NASA Research Park at Moffet Airfield in California.

3 Less Human Warfare = More Humanism?

On the one hand, transhumanist ideas about man and war cannot be dismissed from the outset. It must be allowed to consider whether it is not precisely human passions that repeatedly fuel conflicts as soon as differences of vision and opinion arise. Members of the military apparatus should, on the contrary, behave as professionally as possible, especially soldiers. Professionally means as passionless as possible. As some transhumanists, especially within the military, hope (Wallace 2015), could the increasing use of autonomous weapon systems that is emerging today contribute to enhancing humanistic tendencies in warlike conflicts? Would autonomous (that is intelligent but dispassionate) weapons perhaps bring an era of "clean warfare" that has been hoped for and encouraged for so long, at least since the First Gulf War of 1990/1991? The transhumanist vision of a "self-active" technology that replaces humans, especially in dangerous situations, seems to point in this direction.

Substantially, the transhumanists are indeed concerned with the humanization of conflicts and the military through hyper-intelligent technology, maybe through letting in the future "only" robots or remote-controlled machines fight one another, while humans gradually physically disappear from conflicts and wars. Some transhumanists therefore consider "autonomous" weapon systems as *"friends of man"* (Toscano 2015). This is for example illustrated by the new Russian battle tank T-14 Armata, which also serves as a model for new developments of the West—and scares many by its intelligent automation which leaves only assistance functions to humans. The same is true for the next generation of warplanes and stealth warships, whose "intelligence" is also rapidly increasing, resulting in more and more reconnaissance, identification and combat operations being carried out automatically—and thus faster and (semi-)independently from humans. The same also applies to self-controlling systems, which are still mostly in the experimental stage today. Autonomous and self-driving systems have been the subject of numerous legislative discussions since 2015, also at United Nations level, in order to outlaw or ban them in most contexts.

Although transhumanists are in principle mostly opposed to self-decision-making systems in wars, at the same time they are in favour of a broad and comprehensive civil human–machine convergence. Therein lies a contradiction. The examples are by no means as consistently grim as some suppose, but surely deeply ambivalent. The question is, if, when two robots are done with fighting and one of them survives, does it return home or does it proceed to destroy the people behind the other robot to prevent them from sending a new opponent? The absence of people from war, as desired by transhumanists, would then be only the first step, which had to be followed by the destruction of the people behind the technology, if war were to realize its intrinsic "sense" and goal, namely the generation of a victor and a loser.

4 Are Conflicts Going Extinct Because People Disappear from Them?

The extinction of armed conflicts due to the foreseeable disappearance of people from wars is therefore hardly possible. And the penetration of battlefields with the spirit of humanity is also unlikely. Plundering, rape, abuse and destructiveness are part of the repertoire of warfare, resulting in fear, traumatization, brutalization and hatred of humans in the face of the experience of death that accompanies war. To kill before being killed when in doubt is a determining—human and technical—principle in armed conflicts.

This results in a fundamentally ambivalent perspective regarding the "transhumanist" use of autonomous weapon systems. On the one hand, there is hope for alleviating the suffering of those directly involved in war, that is, fewer dead soldiers (which must be considered logical, if fewer and fewer of them are human) as well as fewer deaths and less suffering on the part of the civilian population due to lack of human passions and of the "*inappropriate behaviour*" deriving from them. For example, this argument is used by Ronald (Ron) Arkin, a professor at the Georgia Institute of Technology and a prominent advocate of the use of deadly autonomous weapon systems (Häusler 2017; Georgia Tech n.d.). The question, however, is whether war is not in and of itself an "inappropriate behaviour".

On the other hand, it is to be feared that less cruelty and passion in war will lead to more wars. The inhibition threshold of combat involvement for military and political decision-makers could decrease if they could expect lower human losses to be called responsible for in the event of a conflict. Harbingers of this development can be found in the fantasies of today's US President Donald Trump to make "small" atom bombs deployable and legitimize their use "ethically" (Wellerstein 2016). To the same extent, public support is likely to increase if "transhumanist" war loses its horror through less human damage. The picture of a clean war thanks to technical accuracy would undoubtedly receive new impetus. But that would be a double-edged sword, which would ultimately turn against humans.

5 Transhumanism and Artificial Intelligence

The assessment of the impact of artificial intelligence (AI), which is perhaps the fastest developing of all avant-garde engineering industries, seems to be the key question in transhumanist circles. In recent years, many intellectuals in the technology industry have repeatedly and publicly expressed their concern in this connection. The list ranges from Steve Wozniak and Bill Gates to Elon Musk and Stephen Hawking. They all see the danger that humans destroy themselves through the creation of artificial intelligence, because artificial intelligence will have to turn against humans for reasons of anticipatory self-preservation: only humans can "pull the plug", and this mere possibility is enough to preemptively eliminate them as a threat factor.

This scenario, described for example in the book *Superintelligence* by Nick Bostrom (2014, p. 117f.), dramatic as it may be in the worst case, and "habitual" as it may be to us thanks to popular culture—think of science fiction imaginations like "Terminator", "Space odyssey", "Westworld" or "Ex Machina"—is perhaps less immediately realistic than we feel to be the case.

Nevertheless, artificial intelligence, which surpasses the intelligence of humans in special cases, has long been omnipresent. From autopilots in planes to chess computers, there is no lack of intelligence in terms of computing and processing. But yet there is (still) a lack of awareness. All the artificial intelligence that exists so far is no more than an illusion of self-conscious consciousness at best—that is, of consciousness that reaches near that of man (the "psychic" stimulus-response reaction is a primordial consciousness that is also present in animals, but without the "me" function and without the "I" experience).[1]

In this regard, the human mind is easily deceived. Wherever we see an indication of intentional action, we assume there is also consciousness. From the stuffed animal to God, this follows the same logic that, according to evolutionary researchers, goes back to an adaptation property called "*agent detection*" (Barrett 2009). A property stemming from the fact that on hearing a rustling bush it is more advantageous for our survival to assume an enemy than a gust of wind. Whether a man-like artificial consciousness can be created and then used for military purposes in the foreseeable future is empirically still uncertain.

6 Human Enhancement and Military-Relevant Technology Innovation

However, there are avant-garde military technologies that already exist in practice. They urgently need public regulations that have been wantonly neglected until now. We are talking about the interface between human enhancement and military technology. What is it about?

One way to establish military superiority through technological innovation is to enhance the equipment and superior performance of human ground forces. For example, the Wyss Institute of Harvard University, in cooperation with the US military research agency DARPA, is investigating exoskeletons, also known as exosuits, to make soldiers perform better, so that they can carry more load over longer distances (Ackerman 2015; Wyss Institute 2017). Combined with this, virtually all world powers also have intrusive technologies on their agenda, such

[1]A pathbreaking example, although by now already somewhat outdated, is the humanoid robot BINA48 (first development 2010) by Hanson Robotics: http://www.hansonrobotics.com/robot/bina48/

as the (experimental) “Cortical Modem”, which in future should allow for feeding information directly into the visual cortex of the brain, so that soldiers in manoeuvres have the data directly in their field of vision (Hewitt 2015), free from external technology. Such a brain–computer interface should, if necessary, also work in the opposite direction and enable the long-term control of robots by means of thought.

The approach consisting in increasing human capacities, whether through exoskeletons, intrusive technologies or the use of behaviour-altering or performance-enhancing pharmaceuticals, already validated for years but hardly covered by military agreements, is unlikely to be fully prevented in the future. This is so, firstly, because human enhancement encompasses a growing range of technological application options, and it is therefore difficult to survey the field, let alone control it comprehensively. Secondly, since it is not clear in advance which of the emerging transhumanist options should be classified as relevant for warfare and to what extent. Many potentially warfare-relevant innovations today come from the medical field and more than ever from the entertainment industry. And it is practically impossible to outlaw research and products in these fields at transnational level because of their possible military usability. This induces another problem, namely that, in addition to state actors, more and more innovations are coming from the transnational private sector, which hinders effective controls.

This applies, for example, to the field of convergence of automation and robotics. The military relevance of this convergence in relation to the use of deadly autonomous weapon systems is certainly less ambiguous than in the case of the principle of human enhancement as such. One such example is the robotics company Boston Dynamics, belonging to the Google empire since 2013. It has been building humanoid robots of the Atlas type since 2012. Like Atlas, the four-legged Cheetah with top speeds of almost 50 km/h, and capable of carrying a load of around 150 kg, will support conventional soldiers in action, but according to the official presentation, it will not replace them for the time being (Boston Dynamics n.d.; Ackerman and Guizzo 2016). Officially, the focus is on the use of rescue operations and other services, not combat deployment. Nevertheless, elements for the development of killer robots could emerge from these prototypes under changed political, legal and international conditions (Krishnan 2016).

7 The Military as a Social Indicator

Conclusion? Nowadays, the basic trend towards more “transhumanist” technology in conflict and war is undoubtedly established. Apart from the legal aspect, how much the future technologies will actually act autonomously and make decisions about life and death is, besides a legal issue, principally a question concerning the definition of autonomy—which is probably the biggest hurdle on the way to outlawing possible killer robots, if desired at all. In practice, the biggest problem for a general prohibition of autonomous weapon systems lies in the fact that the boundary between automatic and autonomous is fluid and a generally valid conceptual clarification is hardly possible. A strict interpretation of this field within the framework of the Geneva Convention can be hardly expected, also because the military interest in new developments and their versatile potential is likely to be too great.

The biggest problem with all of this is not the intentions of the transhumanists, who mostly argue for the banning or at least a strong restriction of autonomous weapon systems (Russell et al. 2015)—which, however, they consider difficult to enforce, as there still is no world government (and such a government cannot be expected for the foreseeable future) that could lighten the prospects of the *"darkness of current politics and ethics"* regarding the confluence of transhumanism and military in a joint and orderly manner (Bostrom 2006). The threat is rather represented by the current international political logic that actually speeds up the technologization of conflict and war. In a world of regressive trends back to nation states, while one is struggling with itself, another one has already advanced one step further. The fear of being left behind accompanies every single decision for or against the use of new technology. This is a logic that quickly leads to an arms race in every area of life, among individuals as well as among states. This certainly also applies in the field of military whose inner logic always gives the compulsion to strive for technological superiority. A strong military interest in the potential of human enhancement in connection with (semi-) autonomous weapon systems is therefore almost self-evident.

Consequently, all major world players invest considerable resources in corresponding "transhumanist" military projects. For example, in 2015 Russia presented to the public its already mentioned new battle tank Armata T-14, which is worldwide regarded as a step towards a new quality of automation and thus as a development model to be followed. In the standard version, the T-14 is equipped with a remote-controlled weapon tower, but in the future, it will also be able to act in a completely automated way as a drone or robot tank (Odrich 2015). According to its manufacturer, the Israeli HARPY drone can autonomously detect enemy radar systems and attack them without explicit orders by virtue of its advanced intelligence and decision-making ability (Israel Aerospace Industries IAI n.d.). The list could continue with any number of examples, since practically all weapon systems are on their way to combine intelligence and "self-activity" (in an expanded and blurred sense).

More clearly than ever before, a global technological imbalance is developing that, in addition to the problem of unequal military means, anticipates and drives a development that is also to be feared globally, even by society as a whole. It is about the big question of our time regarding the growing inequality in means, capabilities and resources, far beyond the military. While one part of humanity transcends the limits of human existence through technology, another, presumably much larger, part lives without access to such opportunities and is still confronted with overly human existential problems such as hunger, poverty, disease and death. This also applies to conflict and war.

8 Outlook: An Ambiguous Perspective

Thus, today the consequence of overzealous belief in technology seems to be the increasing division of humanity into two camps and into a two-speed world. "Transhumanism" may see itself as the heir or even the further development of classical humanism, but with its obstinate insistence on the universal technologization of humans, environment and conflicts, it indirectly contributes to this development, exacerbating the global inequality gap.

However, even considering all the above, transhumanism is not a homogeneous philosophy, and technology itself is neither good nor bad, despite some dubious "transhumanist" euphoria. Rather, it is an extension and complement of human options for action. Its creation, transmission and evolution are maybe the essential element of human evolutionary development. Technology is a tool. And, as in any case of tool use, the result even in conflict situations depends on two factors: first, the ability to use the tool effectively and appropriately and, secondly, the intention and integrity of the users.

So, in the long run, it is probably less critical what technology is available, no matter how scary it may be. What matters more is that it is in our hands, in human hands—and that it is therefore just as ambivalent as the human being itself.

References

Ackerman, E.: DARPA tests battery-powered exoskeletons on real soldiers. IEEE Spectr. http://spectrum.ieee.org/video/robotics/military-robots/darpa-tests-batterypowered-exoskeletons-on-real-soldiers (2015)

Ackerman, E., Guizzo, E.: The next generation of Boston dynamics' ATLAS robot is quiet, robust, and tether free. IEEE Spectr. http://spectrum.ieee.org/automaton/robotics/humanoids/next-generation-of-boston-dynamics-atlas-robot (2016)

Barrett, J.L.: Cognitive science, religion and theology. In: Schloss, J., Murray, M. (eds.) The Believing Primate – Scientific, Philosophical, and Theological Reflections on the Origin of Religion, pp. 76–99. Oxford University Press, New York (2009)

Benedikter, R., Siepmann, K., Reymann, A.: "Head-transplanting" and "mind-uploading" – Philosophical implications and potential social consequences of two medico-scientific Utopias. Rev. Contemp. Philos. **16**, 38–82 (2017). https://doi.org/10.22381/RCP1620172

Boston Dynamics. Changing your idea of what robots can do. http://www.bostondynamics.com/index.html (n.d.)

Bostrom, N.: Technological revolutions – ethics and politics in the dark. In: Cameron, N.M.d.S., Mitchell, M.E. (eds.) Nanoscale: Issues and Perspectives for the Nano Century, pp. 129–152. Hoboken, NJ, Wiley (2006)

Bostrom, N.: Superintelligence – Paths, Dangers. Strategies. Oxford University Press, Oxford (2014)

Georgia Tech. News center – experts guide: Ron Arkin. http://www.news.gatech.edu/expert/ronald-arkin (n.d.)

Global Future 2045 International Congress. Open letter to UN-Secretary General Ban Ki-moon. 12 March 2013. http://gf2045.com/read/208/ (2013)

Häusler, T.: Roboterforscher Ron Arkin – Wie Kampfroboter den Krieg humaner machen sollen. Schweizer Radio und Fernsehen (SRF) – Gesellschaft & Religion. https://www.srf.ch/kultur/gesellschaft-religion/wie-kampfroboter-den-krieg-humaner-machen-sollen (2017)
Hewitt, J.: DARPA dreams – cortical modems and neural RAMplants for restoring active memory. Extreme Tech. https://www.extremetech.com/extreme/203718-darpa-dreams-cortical-modems-and-neural-ramplants-for-restoring-active-memory (2015)
Israel Aerospace Industries IAI. Harpy NG; iai.co.il. https://www.iai.co.il/p/harpy (n.d.)
Istvan, Z.: The Transhumanist Wager. Self-Published, New York (2013)
Krishnan, A.: Killer Robots – Legality and Ethicality of Autonomous Weapons. Routledge, London (2016)
Odrich, P.: Steuerung per Fernbedienung – Russland entwickelt Roboterversion des Panzers Armata T-14. Ingenieur.de. http://www.ingenieur.de/Branchen/Fahrzeugbau/Russland-entwickelt-Roboterversion-Panzers-Armata-T-14 (2015)
Russell, S., Tegmark, M., Walsh, T.: Why we really should ban autonomous weapons. Kurzweil – Accelerating Intelligence. http://www.kurzweilai.net/why-we-really-should-ban-autonomous-weapons-a-response (2015)
Toscano, C.P.: "Friend of humans" – an argument for developing autonomous weapon systems. J. Natl. Secur. Law Policy. **8**(1), 207 (2015)
Wallace, S.: The proposed ban on offensive autonomous weapons is unrealistic and dangerous – so says former U.S. Army Officer and Autonomous Weapons expert Sam Wallace. Kurzweil – Accelerating Intelligence. http://www.kurzweilai.net/the-proposed-ban-on-offensive-autonomous-weapons-is-unrealistic-and-dangerous (2015)
Wellerstein, A.: No one can stop President Trump from using nuclear weapons – That's by design. The Washington Post. https://www.washingtonpost.com/posteverything/wp/2016/12/01/no-one-can-stop-president-trump-from-using-nuclear-weapons-thats-by-design/?utm_term=.9d0b95614293 (2016)
Wyss Institute. Soft Exosuits. wyss.harvard.edu, https://wyss.harvard.edu/technology/soft-exosuit/ (2017)

Part III
Technological Aspects

Chapter 9
The Singularity Hoax: Why Computers Will Never Be More Intelligent than Humans

Adriana Braga and Robert K. Logan

Abstract We argue that the dream of the supporters of the technological singularity, the notion that computers will one day be smarter than their human creators will never be realized. The notion of intelligence that advocates of the technological singularity promote does not take into account the full dimension of human intelligence. Human intelligence as we will show is not based solely on logical operations and computation, but rather includes a long list of other characteristics that are unique to humans that the supporters of the singularity ignore. The list includes curiosity, imagination, intuition, emotions, passion, desires, pleasure, aesthetics, joy, purpose, objectives, goals, telos, values, morality, experience, wisdom, judgement and even humour.

1 Introduction

The notion of the technological singularity or the idea that computers will one day be more intelligent than their human creators has received a lot of attention in recent years. A number of scholars have argued both for and against the idea of a technological singularity using a variety of different arguments. We will show that despite the usefulness of artificial intelligence, the singularity is an overextension of AI and that no computer can ever duplicate the intelligence of a human being because of the many dimensions of human intelligence that involve characteristics that we believe cannot be duplicated by silicon-based forms of intelligence because machines lack a number of essential properties that only a flesh and blood living organism, especially a human, can possess. In short, we believe that artificial

A. Braga
Department of Social Communication, Pontifícia Universidade Católica do Rio de Janeiro (PUC-RJ), Rio de Janeiro, RJ, Brazil

R. K. Logan (✉)
Department of Physics, University of Toronto, Toronto, ON, Canada
e-mail: logan@physics.utoronto.ca

W. Hofkirchner, H.-J. Kreowski (eds.), *Transhumanism: The Proper Guide to a Posthuman Condition or a Dangerous Idea?*, Cognitive Technologies,
https://doi.org/10.1007/978-3-030-56546-6_9

intelligence (AI) or its stronger version artificial general intelligence (AGI) can never rise to the level of human intelligence because computers are not capable of many of the essential characteristics of human intelligence, despite their ability to outperform us as far as logic and computation are concerned. As Einstein once remarked "Logic will get you from A to B. Imagination will take you everywhere".

What motivated us to write this essay is our fear that some who argue for the technological singularity might in fact convince many others to lower the threshold as to what constitutes human intelligence so that it meets the level of machine intelligence, and thus devalue those aspects of human intelligence that we (the authors) hold dear such as imagination, aesthetics, altruism, creativity and wisdom.

To be a fully realized human intelligent being it is necessary, in our opinion, to have these characteristics. We will suggest that these many aspects of the human experience that are associated uniquely with our species *Homo sapiens* (wise humans) do not have analogues in the world of machine intelligence, and that as a result the notion that an artificial intelligent machine-based system that is more intelligent than a human is not possible and that the notion of the technological singularity is basically science fiction. Human intelligence and machine intelligence are of a completely different nature so to claim that one is greater than the other is like comparing the proverbial apples and oranges. They are different and they are both valuable and one should not be mistaken for the other.

There is a subjective, non-rational (or perhaps extra-rational) aspect of human intelligence, which a computer can never duplicate. We do not want to have intelligence as defined by singularitarians, who are primarily AI specialists and as a result are motivated to exaggerate their field of research and their accomplishments as is the case with all specialists. Engineers should not be defining intelligence. Consider the confusion engineers created by defining Shannon's measure of signal transmission as information (see Braga and Logan 2017).

To critique the idea of the singularity we will make use of the ideas of Terrence Deacon (2012), as developed in his study *Incomplete Nature: How Mind Emerged from Matter*. Deacon's basic idea is that for an entity to have sentience or intelligence it must also have a sense of self (ibid., 463–484). In his study, Deacon (ibid., 524) defines information "as about something for something toward some end". As a computer or an AI device has no sense of self (i.e. no one is home), it has no information as defined by Deacon. The AI device only has Shannon information, which has no meaning for itself, i.e. the computer is not aware of what it knows as it deals with one bit of data at a time. We will discover that many of the other critiques of the singularity that we will reference parallel our notion that a machine has no sense of self, no objectives or ends for which it strives and no values.

We will also make use of media ecology and the insights of Marshall McLuhan (1964) that the medium is the message. Our basic thesis is that computers, together with AI, are a form of technology and a medium that extends human intelligence not a form of intelligence itself.

Our critique of AGI will make use of McLuhan's (ibid.) technique of figure/ground analysis, which is at the heart of his iconic one-liner the "medium is the message" that first appeared in his book *Understanding Media*. The medium independent of its content has its own message. The meaning of the content of a medium, the figure, is affected by the ground in which it operates, the medium itself. The problem that the advocates of AGI and the singularity make is they regard the computer as a figure without a ground. As McLuhan once pointed out "logic is figure without ground" (McLuhan 2011). A computer is nothing more than a logic device and hence it is a figure without a ground. A human and the human's intelligence are each a figure with a ground, the ground of experience, emotions, imagination, purpose and all of the other human characteristics that computers cannot possibly duplicate because they have no sense of self.

While we are critical of the notion of the idea of the singularity, we are quite positive regarding the value of AI. We also believe, like Rushkoff (2015, pp. 354–355), that networked computers will increase human intelligence by allowing humans to network and share their insights.

2 The Ground of Intelligence: What Is Missing in Computers

At the core of our critique of the technological singularity is our belief that human intelligence cannot be exceeded by machine intelligence because the following set of human attributes are essential ingredients of human intelligence, and they cannot, in our opinion, be duplicated by a machine. The most important of these is that humans have a sense of self and hence have purpose, objectives, goals and telos, as has been described by Terrence Deacon (2012, pp. 463–484) in his book *Incomplete Nature*. As a result of this sense of self, humans also have curiosity, imagination, intuition, emotions, passion, desires, pleasure, aesthetics, joy, values, morality, experience, wisdom and judgement. All of these attributes are essential elements of or conditions for human intelligence, in our opinion. In a certain sense, they are the ground in which human intelligence operates. Stripped of these qualities as is the case with AI, all that is left of intelligence is logic, a figure without a ground according to McLuhan as we have already mentioned. If those that desire to create a human level of intelligence in a machine, they will have to find a way to duplicate the above list of characteristics that we believe define human intelligence.

To the long list above of human characteristics that we have suggested contributes to human intelligence, we would also add humour. Humour entails thinking out of the box, a key ingredient of human intelligence. Humour specifically works by connecting elements that are not usually connected, as is also the case with creative thinking. All of the super-intelligent people we have known invariably have a great sense of humour. Who can doubt the intelligence of the comics Robin Williams and Woody Allen, or the sense of humour of physicists Albert Einstein and Richard Feynman?

There are computers that can calculate better than us, and in the case of IBM's Big Blue, play chess better than us, but Big Blue is a one-trick pony that is incapable of many of the facets of thinking that we regard as essential for considering someone intelligent. Other examples of computers that exceeded humans in game playing are Google's AlphaGo beating the human Go champion and IBM's Watson beating the TV Jeopardy champion. In the case of Watson, it won the contest, but it had no idea of what the correct answers it gave meant and it did not realize that it won the contest, nor did it celebrate its victory. What kind of intelligence is that? A very specialized and narrow kind for sure.

Perhaps the biggest challenge to our scepticism vis-à-vis the singularity is a recent feat by the non-profit organization OpenAI with the mission of openly sharing its AI research. They developed an AI machine that can play games against itself and thereby find the optimum strategy for winning the game. It played GO against itself for 3 days, and when it was finished, it was able to beat the original AlphaGo computer that had beaten the human Go champion. In fact, it played 100 matches against AlphaGo and it won them all. AI devices that can beat humans at rule-based games parallel the fact that computers can calculate far faster and far better than any human. The other aspect of computers beating humans playing games is that a game is a closed system, whereas life and reality is an open system (Quach 2017).

Intelligence, at least the kind that we value, involves more than rule-based activities and is not limited to closed systems, but operates in open systems. All of the breakthroughs in science and the humanities involve breaking the rules of the previous paradigms in those fields. Einstein's theory of relativity and quantum theory did not follow the rules of classical physics. As for the fine arts, there are no rules. Both the arts and the sciences are open systems.

The idea that a computer can have a level of imagination or wisdom or intuition greater than humans can only be imagined, in our opinion, by someone who is unable to understand the nature of human intelligence. It is not our intention to insult those that have embraced the notion of the technological singularity, but we believe that this fantasy is dangerous and has the potential to mislead the developers of computer technology by setting up a goal that can never be reached, as well as devalue what is truly unique about the human spirit and human intelligence.

It is only if we lower our standards as to what constitutes human intelligence, will computers overtake their human creators as advocates of AGI and the technological singularity suggest. Haim Harari (2015, p. 434) put it very succinctly when he wrote that he was not worried about the world being controlled by thinking machines but rather he was "more concerned about a world led by people who think like machines, a major emerging trend of our digital society". In a similar vein, Devlin (2015, p. 76) claims that computers cannot think, they can only make decisions, and that, he further claims, is the danger of AGI, namely decisions that are made without thought.

3 The 3.5 Billion Year Evolution of Human Intelligence

Many of the shortcomings of AGI as compared to human intelligence are due to the fact that human beings are not just logic machines, but they are flesh and blood organisms that perceive their environment, have emotions, have goals and have the will to live. These capabilities took 3.5 billion years of evolution to create.

We have indicated that human intelligence for us is not just a matter of logic and rationality, but that it also entails explicitly the following characteristics that we will now show are essential to human thought: purpose, objectives, goals, telos, caring, intuition, imagination, humour, emotions, passion, desires, pleasure, aesthetics, joy, curiosity, values, morality, experience, wisdom and judgement. We will now proceed through this list of human characteristics and show how each is an essential component of human intelligence that would be difficult if not impossible to duplicate with a computer. These characteristics arise directly or indirectly because of the fact that humans have a sense of self that motivates these characteristics. Without a sense of self, who is it that has purpose, objectives, goals, telos, caring, intuition, imagination, humour, emotions, passion, desires, pleasure, aesthetics, joy, curiosity, values, morality, experience, wisdom and judgement. How could a machine have any of these characteristics?

Human thinking is not just logical and rational, but it is also intuitive and imaginative. The advocates of the singularity do not take into account that human thought is not just logical and rational, but it is also intuitive, imaginative and even sometimes irrational. Imagination and curiosity are uniquely human and are not mechanical one step at a time. Mechanically trying to program them as a series of logical steps is doomed to fail, since imagination and curiosity defy logic and a computer is bound by logic. Logic can inhibit creativity, imagination and curiosity. Imagination entails the creation of new images, concepts, experiences or sensations in the mind's eye that have never been seen, perceived, experienced or sensed through the senses of sight, hearing and/or touch. Computers do not see, perceive, experience or sense, as do humans, and therefore cannot have imagination. They are constrained by logic and logic is without images.

Another way of describing imagination is to say it represents thinking outside the box. Well, is the box not all that we know and the equivalent ways of representing that knowledge using logic? Logic is merely a set of rules that allows one to show that one set of statements is equivalent to another. One cannot generate new knowledge using logic; one can only find new ways of representing it. Creativity requires imagination and imagination requires creativity and both creativity and imagination are intuitive, so once again we run up against another barrier that prevents computers from generating general intelligence.

Imagination is essential in science for creating a hypothesis to explain observed phenomena, and this part of the process of scientific thinking requires imagination, which is quite independent of logic. Logic comes into play when one used logic to determine the consequences of one's hypotheses that can be tested empirically.

Devising ways to test one's hypotheses requires another, but quite different kind of imagination.

Humans experience a wide variety of emotions, some of which, as Einstein suggests, motivate art and science. Emotions, which are a psychophysical phenomenon, are closely associated with pleasure (or displeasure), passion, desires, motivation, aesthetics and joy. Every human experience is actually emotional. It is a response of the body and the brain. Emotions play an essential part in human thinking which computers are incapable of.

Computers are incapable of emotions, which, in humans, are inextricably linked to pleasure and pain because computers have no pain nor any pleasure, and hence there is nothing to get emotional about. In addition, they have none of the chemical neurotransmitters, which is another reason why computers are incapable of emotions and the drives that are associated with them. Without emotions, computers lack the drive that are an essential part of intelligence and the striving to achieve a purpose, an objective or a goal. Emotions play a key role in curiosity, creativity and aesthetics, which are three other factors that are essential for human intelligence.

Curiosity is both an emotion and a behaviour. Without the emotion of curiosity, the behaviour of curiosity is not possible, and given that computers are not capable of emotions, then they cannot be curious, and hence lack an essential ingredient for intelligence. Curiosity entails the anticipation of reward, which in the brain comes in the form of neurotransmitters like dopamine and serotonin. No such mechanism exists in computers, and hence they totally lack native curiosity. Curiosity, if it exists at all, would have to be programmed into them. In fact, that is exactly what NASA did when it sent its Mars rover, aptly named Curiosity, to explore the surface of Mars.

Without emotions computers cannot experience creativity and aesthetics, two more essential elements of human intelligence.

Because a computer has no purpose, objectives or goals, it cannot have any values as values are related to one's purpose, objectives and goals. As is the case with curiosity, values will have to be programmed into a computer, and hence the morality of the AGI device will be determined by the values that are programmed into it, and hence the morality of the AGI device will be that of its programmers. This gives rise to a conundrum. Whose values will be inputted, and who will make this decision, a critical issue in a democratic society. Not only that, but there is a potential danger. What if a terrorist group or a rogue state were to create or gain control of a super-intelligent computer or robot that could be weaponized. Those doing AGI research cannot take comfort in the notion that they will not divulge their secrets to irresponsible parties. Those that built the first atomic bomb thought that they could keep their technology secret, but the proliferation of nuclear weapons of mass destruction became a reality. When considering how many hackers are operating today, is not the threat of super-intelligent AGI agents a real concern?

Intelligence, artificial or natural, entails making decisions, and making decisions requires having a set of values. So, once again, as was the case with curiosity, the decision-making ability of an AGI device cannot exceed that of human decision-making as it will be the values that are programmed into the machine that will ultimately decide which course of action to take and which decisions are made.

If the challenges of programming an AGI device with a set of values and a moral compass that represents the will of the democratic majority of society are achieved, there is still the challenge of whether the AGI device still has the judgement and wisdom to make the correct decision. In other words, is it possible to program wisdom into a logical device that has no emotions and has no experiences upon which to base a decision. Wisdom is not a question of having the analytic skills to deal with a new situation but rather having a body of experiences to draw upon to guide one's decisions. How does one program experience into a logic machine?

Intelligence requires the ability to calculate or to compute, but the ability to calculate or compute does not necessarily provide the capability to make judgements and decisions unless values are available, which for an AGI device, requires input from a human programmer.

4 Conclusions: How Computers Will Make us Humans Smarter

Douglas Rushkoff (2015, pp. 354–355) invites us to consider computers not as figure but as ground. He suggests that the leap forward in intelligence will not be in AGI-configured computers that have the potential to be smarter than us humans, but in the environment that computers create. Human intelligence will increase by allowing human minds to network and create something greater than what a single human mind can create, or what a small group of minds that are co-located can create. The medieval university made us humans smarter by bringing scholars in contact with each other. The city is another example of a medium that allowed thinkers and innovators to network, and hence increase human intelligence. The printing press had a similar impact. With networked computer technology, a mind with a global scale is emerging.

In the past, schools of thought emerged that represented the thinking of a group or team of scholars. They were named after cities. What is emerging now are schools of thought and teams of scholars that are not city-based but exist on a global scale. An example is that we once talked about the Toronto school of communication and media studies consisting of scholars, such as Harold Innis, Marshall McLuhan, Ted Carpenter, Eric Havelock and Northrop Fry that lived in Toronto and communicated with each other about media and communications. A similar New York school of communication emerged with Chris Nystrom, Jim Carey, John Culkin, Neil Postman and his students at NYU. Today, that tradition lives on but not as the Toronto School or the New York School, but as the Media Ecology School, with participants in every part of the world. This is what Rushkoff (2015) was talking about in his article "The Figure or Ground", where he pointed out that it is the ground or environment that computers create, and not the figure of the computer by itself that will give rise to intelligence greater than a single human. He expressed this idea as follows: "Rather than towards machines that think, I believe we are migrating toward a networked environment in which thinking is no longer an individual activity nor bound by time and space".

Marcelo Gleiser (2015) strikes a similar chord to that of Doug Rushkoff when he points out that many of our technologies act as extensions of who we are. He asks: "What if the future of intelligence is not outside but inside the human brain? I imagine a very different set of issues emerging from the prospect that we might become super-intelligent through the extension of our brainpower by digital technology and beyond—artificially enhanced human intelligence that amplifies the meaning of being human".

References

Braga, A., Logan, R.: Communication, information and pragmatics. In: Khosrow-Pour, M. (ed.) Encyclopedia of Information Science and Technology. IGI Global, Hershey, PA (2017)

Deacon, T.: Incomplete Nature: How Mind Emerged from Matter. W.W. Norton & Company, New York, NY (2012)

Devlin, K.: Leveraging human intelligence. In: Brockman, J. (ed.) What to Think About Machines that Think, pp. 74–76. New York, NY, Harper Perennial (2015)

Gleiser, M.: Welcome to your transhuman self. In: Brockman, J. (ed.) What to Think About Machines that Think, pp. 54–55. New York, NY, Harper Perennial (2015)

Harari, H.: Thinking about people who think like machines. In: Brockman, J. (ed.) What to Think About Machines that Think, p. 434. New York, NY, Harper Perennial (2015)

McLuhan, M.: Understanding Media: Extensions of Man. McGraw Hill, New York, NY (1964)

McLuhan, E.: Media and Formal Cause. NeoPoiesis Press, New York, NY (2011)

Quach, K.: How DeepMind's AlphaGo Zero Learned All by Itself to Trash World Champ AI AlphaGo. https://www.theregister.co.uk/2017/10/18/deepminds_latest_alphago_software_doesnt_need_human_data_to_win (2017)

Rushkoff, D.: Figure or ground? In: Brockman, J. (ed.) What to Think About Machines that Think, pp. 354–355. New York, NY, Harper Perennial (2015)

Chapter 10
Ethical Machine Safety Test

Roman M. Krzanowski and Kamil Trombik

Abstract Within a few decades, autonomous robotic devices, computing machines, autonomous cars, drones and alike will be among us in numbers, forms and roles unimaginable only 20 or 30 years ago. How can we be sure that those machines will not under any circumstances harm us? We need a verification criterion: a test that would verify the autonomous machine's aptitude to make "good" rather than "bad" decisions. This chapter discusses what such a test would consist of. We will call this test the ethical machine safety test or machine safety test (MST) for short. Making "good" or "bad" choices is associated with ethics. By analogy, an ability of the autonomous machines to make such choices is often interpreted as machine's ethical ability, which is not strictly correct. The MST is not intended to prove that machines have reached the level of moral standing people have or reached the level of autonomy that endows them with "moral personality" and makes them responsible for what they do. The MST is intended to verify that autonomous machines are safe to be around us.

1 Introduction

Within the next few decades, autonomous machines will enter our lives not as passive devices but as autonomous agents in unprecedented numbers and roles (Ford 2015; Berg 2016; Bloem et al. 2014; Boyle 2016; Brown 2016; Clifford 2017; Cookson 2016; Krzanowski et al. 2016; Pew Research Center 2014; Schwab 2016, 2017; Sullins 2011). The view that this technology will be all gain and no pain, supported by some,[1] is hardly justified by the historical record and current

[1]"Ethical machines would pose no threat to humanity. On the contrary, they would help us considerably, not just by working for us, but also by showing us how we need to behave if we are to survive as a species" (Anderson and Anderson 2010; Bostrom 2015; Anderson 2016). See also Schwab (2016, 2017).

R. M. Krzanowski · K. Trombik (✉)
The Pontifical University of John Paul II, Cracow, Poland

W. Hofkirchner, H.-J. Kreowski (eds.), *Transhumanism: The Proper Guide to a Posthuman Condition or a Dangerous Idea?*, Cognitive Technologies,
https://doi.org/10.1007/978-3-030-56546-6_10

experience (see Heron and Belfort 2015). A more cautious approach, such as that suggested by Ford (2015), Gray (2007), Yampolskiy (2012a, b) and Kaczynski (1995), is preferable.

The question, then, is the following: How can we be sure that autonomous machines will not harm us and will behave as "ethical" agents?[2] To address this problem we need a test that would verify the autonomous machine's aptitude to act in a safe way, i.e. in some sense to act ethically. We will call such a test the machine safety test (MST).[3]

2 Key Terms

To avoid any misinterpretations, let us define certain key terms used in this chapter. These are ethics, ethical agent, machine ethics, autonomous machines and autonomous ethical agents.

Ethics, in the most general terms, is a set of prescriptions or rules about how to live a good and rewarding life as an individual and as a member of society (Bourke 2008; Vardy and Grosch 1999; MacIntyre 1998). Such a concept of ethics may be reduced, as is often the case, to a set of rules with a yes/no answer, specifying what to do (Beavers 2011). But ethics is more than just rules. It (implicitly or explicitly) requires free will, consciousness, a concept of good and bad, virtue and values, a concept of self, an understanding of responsibility (of "ought" and "ought not") (MacIntyre 1998; Veach 1973; Sandel 2010) and a good comprehension of the reality around us. A lot of deep metaphysics (free will, a concept of good, a concept of self, etc.) is involved in the concept of ethics. Dispensing with metaphysics leaves ethical statements groundless.

[2]We need to keep in mind that the future full of happiness and unalloyed human flourishing promised by light-minded AI and robotic enthusiasts is just an uncritical and hardly justified fairy tale fantasy. I propose to leave behind Start Trekfans, Asimov's Three Laws of Robotics and other Sci-Fi phantasms. History does not justify such a vision at all (unfortunately!). Recall the cautionary words about progress offered by more discerning minds: "What the Enlightenment thinkers never envisioned was that irrationality would continue to flourish alongside rapid development in science and technology… In fact, [there is] no consistent link between the adoption of modern science and technology on the one hand and the progress of reason in human affairs on the other … There is nothing in the spread of new technologies that regularly leads to the adoption of what we like to think of as a modern, rational worldview" (Gray 2007, p. 18).

[3]"The development of machines with enough intelligence to assess the effects of their actions on sentient beings and act accordingly may ultimately be the most important task faced by the designers of artificially intelligent automata" (Allen et al. 2000). Seibt writes: "…we are currently in a situation of epistemic uncertainty where we still lack predictive knowledge about the individual and socio-cultural impact of the placement of social robots into human interactions space, and we are unclear on which aspects of human interactions with social robots lend themselves to predictive analysis" (Seibt 2012).

Can then such a deep ethics be computed (in the Church–Turing sense), given that metaphysics is not mathematical? Ethical rule-based on Hobbesian, Kantian, utilitarian or other ethical schools can be to some extent translated into a computer algorithm and made "computable". But then all "metaphysical" dimensions of the ethical actor are "lost in translation". If a machine is programmed according to "translated" rules, one may claim that it possesses ethical qualities or that it is an ethical machine (Anderson and Anderson 2010). But this ethics would be a special type of ethics, not ethics in the deep, metaphysical sense. Ethics in a deep sense (like metaphysics) is non-computable, and we do not have any other meaning of "computable" that could rescue "computerized ethics" from its shallows (Yampolskiy 2012a, b; Krzanowski et al. 2016)[4] (see Turner 2016 for a definition of computation). The ethical rules translated into a machine format constitute what we will call machine ethics as m-ethics.[5]

An ethical agent is an individual (artefact or natural) acting according to ethical rules. In the context of our discussion, we may call an autonomous machine an ethical agent understanding that we mean here ethics as m-ethics, or a set of behavioural rules directing the behaviour of an autonomous machine. Nothing else. Thus, it is misleading to talk about "moral machines", "ethical machines" or the likes. Too generous use of these terms will only confuse the problems we face with autonomous machines (by attributing to them capacities they cannot have); we need to constantly remind ourselves that the subject of our discussion is autonomous machines with implemented m-ethical software or system.

Autonomous machines (e.g. Floreano et al. 1998; Patrick et al. 2008; Ni and Leug 2016) are machines that act in the environment without direct command by or involvement with humans. Autonomous machines, which implement m-ethics rules and thus display ethical-like behaviours, are autonomous ethical machines.

3 Where We Are with MST

So far we have only a few proposals of such a test. These are:

[4]For someone that cannot accept a concept of deep ethics, the more technical explanation of what ethics really entails may be easier to comprehend: "... given the complexity of human values, specifying a single desirable value is insufficient to guarantee an outcome positive for humans. Outcomes in which a single value is highly optimized while other values are neglected tend to be disastrous for humanity, as for example one in which a happiness-maximizer turns humans into passive recipients of an electrical feed into pleasure centers of the brain. For a positive outcome, it is necessary to define a goal system that takes into account the entire ensemble of human values simultaneously" (Yampolskiy and Fox 2012).

[5]It is critical to understand this difference. If we attribute ethics to machines we may be tempted to bestow on them personality, responsibility and similar (which unfortunately is slowly happening). But if we say that these machines have m-ethics, which is what they have, we will make such flights of fancy much more difficult.

1. Moral Turing test (MTT)
2. Turing triage test (TTT)
3. Ethical competence test (ECT)

The moral Turing test (MTT) proposed by Allen et al. (2000) is similar to the "imitation game" proposed by Turing (1950). In the original Turing formulation of the "imitation game", machine "intelligence" is assessed by a panel of judges based on a series of responses to questions posed to a machine and a human subject. If, based on the answers given, the judges cannot distinguish a machine from a human, then the machine has passed the test, but what exactly it means is open to interpretations as pointed out for example by Oppy and Dove (2016). The MTT would be run in a similar way, but with the key difference that the questions would be of moral import. Allen, Vermer and Zinser recognized the multifarious nature of ethical problems that would beset such a test and thus recognized its potential limitations. The problem with the MTT, however, may lie elsewhere. The Turing test has not lived up to its promise or its author's intentions, and there is no consensus as to what the TT is actually testing or attempting to demonstrate (Turing 1950; Oppy and Dove 2016; Saygin et al. 2000). Thus, if the TT is not clear regarding its meaning,[6] on what grounds can it be extended to ethical problems with any expectation of success?

Another proposal to apply the TT to an ethical machine test is the Turing triage test (TTT) described by Sparrow (2004, 2014). In the test, the AI-based machine (robot, autonomous artificial agent), following an electrical shortage in a hospital, must choose between (1) turning itself off (equivalent to a human suicide in some respects) and saving a patient and (2) saving itself and allowing a patient to die. Sparrow claims that:

> My thesis, then, is that machines will have achieved the moral status of persons when this second choice has the same character as the first one. That is, when it is a moral dilemma of roughly the same difficulty. For the second decision to be a dilemma it must be that there are good grounds for making it either way. It must be the case therefore that it is sometimes legitimate to choose to preserve the existence of the machine over the life of the human being. These two scenarios, along with the question of whether the second has the same character as the first, make up the Turing Triage Test (Sparrow 2004).

Sparrow himself suggests that a machine could never pass the TTT; thus, a machine could never achieve the moral status of a person. Apart from the fact that the TTT has little to do with the original TT (only that a test is applied to a computing machine), it is within that class of abstract ethical problems that includes the notorious trolley problem, terror bomber, strategic bomber or similar (keeping in mind the differences). The "imaginary ethical problems"(also known as thought

[6]The objective of the Turing test (TT) was not to verify some specific kind of intelligence; it was aimed at a general intelligence. Thus, success in playing Chess or Go did not in fact prove or disprove a machine's capacity to reason, according to the TT's requirements.

experiments) have been devised not to provide a test of general ethical abilities[7] but to expose the multifarious nature of ethics. Thus, solving of the Turing triage problem will not translate itself into solving a problem of mariners stranded at sea (Sandel 2010); similarly, an ability to excel in strategic military games does not guarantee success in leading a real battle (as examples from history testify). Complex abstract ethical puzzles tell us little about the solutions to practical ethical problems (Dancy 1985; Szabó 2000; Elster 2011; Lehtonen 2012; Cathcart 2013).

A different proposal, not based on the Turing test, came from Moor (2006). Moor suggested the ethical competence test (ECP), which would assess the ethical aptitude of the computing agent using hypothetical situational scenarios (but not of the "imaginary ethical paradoxes" type). Such scenarios often do not have yes/no options, but instead use less or more favourable ones. The responses of an artificial agent to such scenarios would be compared with the choices made by humans in the comparable situations. Moor also introduced a requirement that the ethical robot provide a defensible and convincing justification for its decision. As he writes: "If a robot could give persuasive justifications for ethical decisions comparable to or better than those of good human ethical decision-makers, then the robot's competence would be inductively established for that area of ethical decision-making" (Moor 2009). As he points out, ethical tests should be situation-dependent as an automated agent may be competent in some situations and yet not in others. In certain situations, Moor points out, computer agents, because of their huge information processing powers, can make better (and faster) decisions than human agents regarding, for example, the allocation of scarce resources or scheduling the delivery of supplies in the event of catastrophic situations to avoid waste. But these decisions would qualify rather as better or optimal managerial decisions, rather than as ethical ones per se. Moor's proposal warrants attention as it acknowledges the complexity of ethical decisions, their dependence on situational context and the need to view an artificial agent not as a moral black box but, at the very least, as a grey one.

Summing up current efforts on testing of m-ethics, we can say that no conclusive proposal is on the table. We do not have a test that could serve as the MST of the general m-ethical capabilities of an artificial agent, nor do we have (implemented, proposed or conceptualized) a testing methodology to construct such a test. As well, we do not know what it would take to test an autonomous ethical agent for its ethical (m-ethical) probity.

[7]"Imaginary stories and thought experiments are often used in philosophy to clarify, exemplify, and provide evidence or counterevidence for abstract ideas and principles. Stories and thought experiments can illustrate abstract ideas and can test their credibility, or, at least, so it is claimed. As a by-product, stories and thought experiments bring literary, and even entertaining, elements into philosophy" (Lehtonen 2012).

4 Machine Safety Test: What It Should Be?

We take a safe approach to the MST (advocated by Yampolskiy 2012a, b; Moor 2009) and formulate three assumptions that underline the definition of the MST:

1. The inherent complexity of ethics renders it incomputable in the TM sense.
2. Computing machines cannot be ethical in the way people are.
3. Computing machines may play the ethical game. It just means that autonomous machines/agents may be made to act amicably towards us in all foreseeable situations and they should be safe to be around.

5 Claim: What We Are Testing

We need to ask: What are the objectives of the MST? Are we testing whether autonomous machines behave like humans or like ethical artificial agents with m-ethics capabilities? The first case is impossible to achieve considering our definition of m-ethics. The machine safety test is NOT the test verifying the *general "moral" or ethical aptitude of a machine*. It means that we do not test the machine's "equivalence" to a human agent, and it also means that we do not test a moral aptitude in specific circumstances.

The machine safety test is not designed to prove that machines have reached the same level of moral standing as people or have reached the level of autonomy that endows them with "moral personality" and makes them responsible for what they do.[8] The objective of the test is only to verify that:

1. The product we develop, i.e. an autonomous agent, a machine, is "safe" to be around people in general circumstances.
2. The propensity of the autonomous machine to do harm, by "intention", design or neglect, is limited to as narrow a margin as reasonably possible. However, it seems that the possibility of doing harm cannot be completely eliminated.

Of course, terms such as "general circumstances", "safe" and "a narrow margin" can be interpreted in many ways. Here, I use common understandings of these terms, accepting that it may require further elaboration. It is possible that soon we will have to create a new dictionary of ethical terms that would correctly represent machine ethics, as current ethical terms may not be sufficient to describe the complex human–machine interactions.

[8]We want to avoid dilemmas as reported by Heron and Belfort (2015): "The question of who we should blame when a robot kills a human has recently become somewhat more pressing. The recent death (we talk about 2015) of a Volkswagen employee at the hand of an industrial factory robot has left ethicists and legislators unsure of where the moral and ethical responsibility for the death should lie—does it lie with the owners, the developers, the factory managers, or elsewhere?"

6 What Should the Test Be?

It seems that the MST should not be theoretical or primarily theoretical (theoretical meaning involving only ethical reasoning verified in a dialogue or a conversation—free or structured). Theoretical questioning is not a good test of moral aptitude. As we know, behind the clever and reasonable answers there may be nothing of substance but software, nothing of substance in an ethical sense, as the experience with chat bots (and politicians) teaches us. And besides, what kinds of questions would we ask? Certainly, any type of standard personality test or tests that try to gauge a person's sanity should be ruled out as the responses to them can be easily programmed and reproduced by a computing machine without even the smallest ethical insight. Imaginary ethical cases as proposed in the TTT are not suited to this function, as explained earlier. Yet, there may still be some use for the verbal verification of the moral standing. Such a verbal test in the form of an interview (qualifying conversation?) may be used to understand the ethical reasoning of a machine (correctness of the software implementation of ethical capacities?), and it may be more conducive to the purpose of the MST than a test requiring specific answers. The requirement that an autonomous ethical agent be able to explain itself stipulates a requirement that it be implemented as a white box. The problem as to how we would gauge the results of such an interview remains an open question.

The bulk of the MST test should be an ability to make just decisions in specific life situations. Making just decisions in a real-life context, not an abstract ability to assign "right" or "wrong" labels to abstract situations, seems to be at the core of ethics.[9] Of course, to act as an ethical agent in life situations an agent will have to possess considerable abstract knowledge of ethics. But we would rather require that an autonomous machine makes ethical choices in concrete situations rather than be able to respond to complex ethical questions or solve imaginary cases.

We may compare the MST to the skipper patent test or an airline pilot test ceteris paribus. These tests include theoretical and practical components. Learning or testing for a pilot or a skipper begins with theoretical tests of basic technical knowledge, before progressing through training on the flight simulators and then flights with an instructor. Finally, these tests also include a period of apprenticeship. In the case of an airline pilot, the pilot-to-be must fly in a junior position for a certain number of hours, before he or she can be recognized as a pilot[10] with the licence to undertake solo flights, likewise with a skipper permit. In the case of the MST, the

[9]"'Ethics', as understood in modernity, focuses on the rightness and wrongness of actions. The focus is misleading in that actions never occur outside of the wider social and natural contexts to which they respond. Individual, community, and society clearly constitute such contexts, on the different levels of the natural ... 'human' world. This world comprises our interpersonal relationships as well as the natural givens" (McCumber 2007, p. 161).

[10]See, for example, the requirements for the testing standards for an airline pilot: https://www.faa.gov/ training_testing/testing/ test_standards/media/faa-s-8081-20.pdf

tests are even more complex than for a specific job function, as they will be testing more general situations.

Thus, it seems that the proper test of the artificial moral agent should consist of a theoretical part, a series of practical problems of varying scope and difficulty, progressing from staged scenarios through to gradually less controlled situations and ending up with completely uncontrolled life situations, and (maybe) include a period of apprenticeship during which we verify an agent's capacity to think and act morally in real-life situations (we would call such situations "open-ended").

Thus, in summing up the discussion, the MST should include the following components:

1. Theoretical verification of ethical aptitudes and reasoning—possibly a qualifying interview rather than a Q&A session plus a white box option.
2. A situational test or series of tests, in which an artificial agent makes autonomous decisions in the fully life-like (controlled or not) environment. The tests may have a different scope and increasing complexity and include:
   - Staged tests
   - Controlled life situations
   - Open-ended situations
3. A period of apprenticeship in which an artificial agent acts in the real conditions under close supervision.

7 Use Case Framework for the MST

Due to the generality of the MST test, only the high-level framework of the UC could be provided. This would consist, in the proper sequence, of four stages:

1. Interview and discussion that would verify understanding of m-ethical rules and m-ethical reasoning using imaginary ethical cases. The tests should be performed under a white box paradigm; i.e. the tested system should be able to explain its decisions
2. Staged situational tests that would verify an ability of a tested system to respond to complex (arranged) situations. These tests may be similar to those used on human subjects such as Milgram experiment (Milgram 1963), Phone booth and Dime experiment (Doris 2002), Stanford Prison Experiment (2014) or Cornell experiment (Doris 2002).
3. Situational tests including controlled and open-ended life situations that would verify an ability of a tested system to respond to complex life situations. In this case, any real-life situation of substantial ethical import could be used, in particular situations prone to dilemmas and conflicts.
4. Apprenticeship, which would test an autonomous machine's ability to act without supervision in real-life environment by participation in real situations.

The white box paradigm, as it was pointed out in the Stage 1, applies to all four stages of testing.

8 Operational Concerns

It is not obvious how the MST should be implemented. It obviously requires a machine capable of human-like functioning. A hardware-embedded software would be incapable of situational tests or apprenticeship without significantly compromising the MST framework. Thus, such devices by definition would not qualify as ethical agents and would not be subjected to the MST.

The learning period for an ethical agent would be long; yet because of the nature of computer technology, it may be that only selected exemplars of robots would be tested and the gains in ethical aptitude may be shared by appropriate updates within the compatible class of machines. Thus, there is no need to test every machine; only selected units should be tested, and the experience would be passed onto other agents.

The learning process for autonomous robots is not well defined. How the autonomous system would learn the proper responses to complex situations and how these responses would be integrated into their m-ethical data base is not clear. This should be another area of research.

It seems also that computing technology would allow the ethical experience to be "inherited" from generation to generation of ethical machines, provided that ethical norms stay unchanged. Thus, the ethical testing would not have to be done *ab ovo* with each new version of machines, something that we humans cannot avoid.

9 Review and Summary

It seems that the MST should include several testing venues, leaning mostly towards solving practical life situations. Such complex tests would verify the ability to make ethical decisions in the presence of conflicting cognitive stimuli, conflicting values and time pressure. Table 10.1 below shows possible components of such a test. The elements of the MST are arranged from the most elementary (Level I) and as such of lower importance to the most critical and complex (Level IV) in the rising degree of importance.

We should ask the question whether every autonomous agent should pass all of these test levels, or maybe, we could accept different levels of "robot ethics" and accept after Seibt (2017) "partial realizations" as applied to the MST, depending on the robot design?

Table 10.1 Proposed structure and components of the machine safety test

	Level	Test component	Objectives	Possible implementation
Theoretical component	I	Interview and discussion	Verify understanding of ethical rules and ethical reasoning	Imaginary ethical cases, a white box paradigm for ethical decisions
Practical component	II	Situational tests—staged	Test ability to respond to complex (arranged) situations	Milgram experiment (Milgram 1963) Phone booth and Dime experiment (Doris 2002) Stanford Prison Experiment (2014) Cornell experiment (Doris 2002)
	III	Situational tests—from controlled to open-ended life situations	Test ability to respond to complex life situations	Any real-life situation of substantial ethical import
	IV	Apprenticeship	Test ability to act without supervision in real-life environment	Participation in real situations—war relief effort, etc.

10 How Would We Evaluate Results?

How would we know that the machine passed the test? One option is to have a panel of judges to review the results of the test and develop test-passing criteria as in the Turing proposal. Should we also accept the Turing criterion of a 70% pass score? If we do, what would it mean to have a 70% ethical agent? Or, would we accept a 70% ethical machine to be among us? It is easier, it seems, to use a 70% pass score to judge that a machine functions reasonably (this was a Turing proposal), but not whether it has a 70% moral aptitude. It seems that any number, short of 100%, as the criterion of acceptance of the MET results would be, in this case, an arguable qualification. But how are we to judge situational tests?

We need to admit that we are not sure how the MST should be graded and what it would mean for the autonomous agent to pass/fail the test.

Perhaps instead of a numerical score, we ought to assign some qualitative "moral" standing to ethical machines. Moor (2009) proposed four classes of ethical robots, or as he calls them—ethical impact agents. These are unethical agents, implicit ethical agents, explicit ethical agents and full ethical agents. These are interesting classifications of hardware–software constructs. However, the four classes are too crude to address the ethical capacity of autonomous robots required by the MST. The best we can say is that these classes mark the points on the spectrum of ethical aptitude from inert objects to human agents, but the scale by nature admits fuzzyfied, not crisp, classes.

Some suggestions as to the gradation of ethical abilities may come from HRI research. Seibt (2017) proposes in the context of human–robot interactions "five notions of simulation or partial realization, formally defined in terms of relationships between process systems (approximating, displaying, mimicking, imitating, and replicating)". With the MST assumption that ethical machines "play an ethical imitation game" (not in the Turing sense of the game) and do not replicate human ethical abilities, such a classification may help us in understanding and classifying the MST results.

11 Parting Comments

We cannot exclude the possibility that the meaning of ethics or morality will evolve to the point that in the future ethical or moral principles attributed to humanity would be attributable to machines, robots, software or the like. Meanings of the words do evolve. Yet it is and will be important to make sure that now and in the future "machine ethics" means behavioural rules for machines, or machine safety specifications, not ethics in the human context. And the MST is supposed to test just this, not the presence of some kind of metaphysical moral fibre in hardware or software.

It seems that one of the barriers in the conceptualization of the MST is that ethical agents are perceived as "computers" or software bundles, in the same way as Turing conceptualized the Turing test subject (which is why we have TTT and MTT proposals). The ethical agents will be machines that act, move, interact with us in physical, not only mental, space. A small taste of this is offered by self-driving cars, which are essentially tested as the MST test is structured including software development, driving in a controlled environment,[11] driving with a supervisor and autonomous driving[12] (Stillgoe 2017a; Hern 2017; Balch 2017). These are essentially four stages of the MST. In the context of self-driving cars, these tests are called social learning (Stillgoe 2017b).

It is rather difficult to imagine that machines will have the same complex of values that people have and thus the same responsibilities towards us. Thus, the MST will verify not how close computing machines come to us, but rather how close they come to our expectations about safe, autonomous machines. One must consider that our expectations regarding autonomous machines will evolve. Another challenge

[11]"Michigan is also home to 'M City,' a 23-acre mini-city at the University of Michigan built for testing driverless car technology". Available at: http://fortune.com/2017/01/20/self-driving-test-sites/

[12]"The carmaker's autonomous vehicles traveled a total of 550 miles on California public roads in October and November 2016 and reported 182 'disengagements,' or episodes when a human driver needs to take control to avoid an accident or respond to technical problems, according to a filing with the California Department of Motor Vehicles. That's 0.33 disengagements per autonomous mile. Tesla reported that there were 'no emergencies, accidents or collisions.' Tesla's report for 2015 specified that it didn't have any disengagements to report" (Hall 2017).

will be posed by the fact that the machine technology has global reach, while m-ethics (as any ethics) is quite often local. Thus, training of "ethical" machines will have to keep pace with these changes; otherwise we may risk meeting on our streets autonomous agents with behavioural propensities of cavemen.

And above all, we need to constantly keep in the mind the fact that we are training or developing machines to act safely, to not to harm us—this is the essence of m-ethics. If, by some fit of imagination, we call it ethical training so be it, as long as we are aware of the difference—thus, no moral robots, no ethical robots, just safely operating autonomous machines.

If history teaches us anything, in this case it may indicate that the ethics of autonomous artificial agents may go the same way as Internet security or software in general: just as software companies do not take responsibility for damage caused by their faulty software, so they will shed the responsibility for the transgressions of their faulty ethical agents. Thus, willingly or not, we may have to learn how to live with Microsoft-Windows-quality ethical machines.[13] Because is there any reason why the future should be any different? It rarely is; it just presents itself in different technology.

Acknowledgments We would like to thank Prof. Pawel Polak for his constructive comments on the early draft of this paper. All the errors, faulty conclusions and logical and factual mistakes are of course of our doing.

References

Allen, C., Varner, G., Zinser, J.: Prolegomena to any future artificial moral agent. J. Exp. Theor. Artif. Intell. **12**, 251–261 (2000)

Anderson, S.: The promise of ethical machines. https://www.project-syndicate.org/commentary/ethics-for-advanced-robots-by-susan-leigh-anderson-2016-12 (2016)

Anderson, M., Anderson, S.L.: Robot be good. Sci. Am. **10**, 72–77 (2010)

Balch, O.: Driverless cars will make our roads safer, says Oxbotica co-founder. https://www.theguardian.com/sustainable-business/2017/apr/13/driverless-cars-will-make-our-roads-safer-says-oxbotica-co-founder (2017)

Beavers, A.F.: Is ethics computable. Presidential Address, Aarhus, Denmark, July 4. http://www.afbeavers.net/cv (2011)

Berg, A.: Revolution evolution. Finance Dev. (2016)

[13]A few quotations substantiate this claim: "Microsoft likes to have everything glued together like a kindergarten art project gone berserk, but this is ridiculous" (Vaughan-Nichols 2014); "*Microsoft Windows isn't the only operating system for personal computers, or even the best . . . it's just the best-distributed. Its inconsistent behavior and an interface that changes radically with every version are the main reasons people find computers difficult to use. Microsoft adds new bells and whistles in each release and claims that this time they've solved the countless problems in the previous versions . . . but the hype is never really fulfilled*" (Anonymous, available at: http://alternatives.rzero.com/os.html [Accessed on 5/1/2017]).

Bloem, J., van Doorn, M., Duivestein, S., Excoffier, D., van Maas, R. Ommeren, E.: Fourth industrial revolution. VINT research report. 3 of 4. https://slidelegend.com/queue/the-fourth-industrial-revolution-sogeti_59b503731723ddf2725f00c7.html (2014)
Bostrom, N.: Superintelligence. Oxford University Press, Oxford (2015)
Bourke, V.J.: History of Ethics, Vol. I, V.II. Axios Press, Mount Jackson, VA (2008)
Boyle, A.: AI prophets say robots could spark unemployment – and a revolution. Geekwire, February 13 (2016)
Brown, A.: YOUR job won't exist in 20 years: Robots and AI to 'eliminate' ALL human workers by 2036. https://www.express.co.uk/life-style/science-technology/640744/Jobless-Future-Robots-Artificial-Intelligence-Vivek-Wadhwa (2016)
Cathcart, T.: The Trolley Problem. Workman Publishing, New York (2013)
Clifford. C: The robots will take our jobs. Here's why futurist ray Kurzweil isn't worried. Entrepreneur. https://www.entrepreneur.com/article/272212 (2017)
Cookson, C.: AI and robots threaten to unleash mass unemployment, scientists warn. Financial Times. February (2016)
Dancy, J.: The role of imaginary cases in ethics. Pac. Philos. Q. **66**, 141–153 (1985)
Doris, J.M.: Lack of Character: Personality and Moral Behavior. Cambridge University Press, Cambridge (2002)
Elster, J.: How outlandish can imaginary cases be? J. Appl. Philos. **28**(3), 2011 (2011)
Floreano, D., Godjecac, J., Martinoli, F., Nicoud, J.-D.: Design, control and applications of autonomous mobile robots. Swiss Federal Institute of Technology, Lausanne. https://infoscience.epfl.ch/record/63893/files/aias.pdf (1998)
Ford, M.: The Rise of Robots: Technology and the Threat of Jobless Future. Basic Books, New York (2015)
Gray, G.: Heresies Against Progress and Other Illusions. Granta Publications, London (2007)
Hall, D.: Tesla Is Testing Self-Driving Cars on California Roads. https://www.bloomberg.com/news/articles/2017-02-01/tesla-is-testing-self-driving-cars-on-california-roads (2017)
Hern A.: Google's Waymo invites members of public to trial self-driving vehicles. https://www.theguardian.com/technology/2017/apr/25/google-self-driving-waymo-invites-members-public-trial-vehicles-phoenix-arizona (2017)
Heron, M., Belfort, P.: Fuzzy ethics: or how I learned to stop worrying and love the bot. SIGCAS Comput. Soc. **45**(4), 13 (2015)
Kaczynski, T.: Industrial society and its future. http://editions-hache.com/essais/pdf/kaczynski2.pdf (1995)
Krzanowski, R. Mamak, K. Trombik, K., Gradzka, E.: Ethics computable, non-computable or nonsensical? In: Defense of Computing Machines. Machine Ethics and Machine Law Conference. Jagiellonian University, Cracow, Poland, 18–19 November 2016
Lehtonen, T.: Idealization and exemplification as tools of philosophy. E-logos. Electro. J. Philos. **16** (2012)
MacIntyre, A.: A Short History of Ethics. Notre Dame Press, Notre Dame (1998)
McCumber, J.: Reshaping Reason. Indiana University Press, Bloomington (2007)
Milgram, S.: Behavioral study of obedience. J. Abnorm. Soc. Psychol. **67**(4), 371–378 (1963)
Moor, J.H.: The nature, importance and difficulty of machine ethics. IEEE Intell. Syst. 18–21 July/August 2006
Moor, J.H.: Four kinds of ethical robots. Philosophy Now. **72**, 12–14 (2009)
Ni, R., Leug. J.: Safety and liability of autonomous vehicle technologies. https://groups.csail.mit.edu/mac/classes/6.805/student-papers/fall14-papers/Autonomous_Vehicle_Technologies.pdf (2016)
Oppy, G., Dove D.: The Turing Test. The Spring 2016 Edition of the Stanford Encyclopedia of Philosophy. http://plato.stanford.edu/archives/spr2016/entries/turing-test/ (2016)
Patrick, L., Bekey, G., Abney, K.: Autonomous Military Robotics: Risk, Ethics, Design. US Department of Navy, Office of Naval Research, Arlington (2008)

Pew Research Center. AI, robotics, and the future of jobs. http://www.pewinternet.org/2014/08/06/future-of-jobs/ (2014)
Sandel, M.J.: Justice: What's the Right Thing to Do? Penguin Books, London (2010)
Saygin, A.P., Cycelki, I., Akman, V.: Turing test: 50 years after. Mind. Mach. **10**, 463–518 (2000)
Schwab, K.: Why everyone must get ready for the 4th industrial revolution. http://www.forbes.com/sites/bernardmarr/2016/04/05/why-everyone-must-get-ready-for-4th-industrial-revolution/2/#a9fc30f40c8c (2016)
Schwab, K.: The Fourth Industrial Revolution. Crown Business, New York (2017)
Seibt, J.: "Integrative social robotics" - a new method paradigm to solve the description problem and the regulation problem? In: Frontiers in Artificial Intelligence and Applications. Volume 290: What Social Robots Can and Should Do. IOS Press, Amsterdam (2012)
Seibt, J.: Towards an ontology of simulated social interaction. In: Hakli, R., Seibt, J. (eds.) Sociality and Normativity for Robots. Studies in the Philosophy of Sociality, vol. 9. Springer, New York (2017)
Sparrow, R.: The Turing triage test. Ethics Inf. Technol. **6**(4), 203–213 (2004)
Sparrow, R.: The Turing Triage Test. When is a robot worthy of moral respect? http://www.thecritique.com/articles/the-turing-triage-test-when-is-a-robot-worthy-of-moral-respect/ (2014)
Stanford Prison Experiment. https://www.prisonexp.org (2014)
Stillgoe, J.: Self-driving cars will only work when we accept autonomy is a myth. https://www.theguardian.com/science/political-science/2017/apr/07/autonomous-vehicles-will-only-work-when-they-stop-pretending-to-be-autonomous (2017a)
Stillgoe, J.: Machine learning, social learning and the governance of self-driving cars. https://papers.ssrn.com/sol3/papers.cfm?abstract_id=2937316 (2017b)
Sullins, J.: Introduction: open questions in roboethics. Philos. Technol. **24**, 233 (2011)
Szabó, G.T.: Thought Experiment: On the Powers and Limits of Imaginary Cases. Routledge, New York (2000)
Turing, A.M.: Computing machinery and intelligence. Mind. **49**, 433–460 (1950)
Turner, R.: The Philosophy of Computer Science. The Winter 2016 Edition of the Stanford Encyclopedia of Philosophy. https://plato.stanford.edu/entries/computer-science/ (2016)
Vaughan-Nichols, S.J.: At Microsoft, quality seems to be job none. Computerword. 16 December 2014
Vardy, P., Grosch, P.: The Puzzle of Ethics. Fount, London (1999)
Veach, H.B.: Rational Man. Indiana University Press, London (1973)
Yampolskiy, R.V.: Leakproofing singularity - artificial intelligence confinement problem. J. Conscious. Stud. (JCS). **19**(1–2), 194 (2012a)
Yampolskiy, R.V.: Artificial intelligence safety engineering: why machine ethics is a wrong approach. In: Müller, V.C. (ed.) Philosophy and Theory of Artificial Intelligence, SAPERE, vol. 5, pp. 389–396. Springer, New York (2012b)
Yampolskiy, R.V., Fox, J.: Safety engineering for artificial general intelligence. Topoi. https://intelligence.org/files/SafetyEngineering.pdf (2012)

Chapter 11
"Action" and Ascription: On Misleading Metaphors in the Debate About Artificial Intelligence and Transhumanism

Rainer Rehak

Abstract In all areas of science communication, metaphors are a very important tool for explaining complex matters in everyday language. While in mathematics or physics, the domain language is visible as such, the notions used in the interdisciplinary field of artificial intelligence (AI) already seem like everyday language and therefore hide their inner complexity. AI vocabulary comprises many anthropomorphisms to describe seemingly cognitive-like computer functions. Considering terms such as "learning", "deciding", "recognizing" or even "intelligence" itself, purely descriptive conversations about the field are nearly impossible. Difficulties now arise when such charged vocabulary as in the field of AI is being directly transferred into other domains or back into everyday language used in political or public debates. Confusion expands even more, when such assumed capabilities hit the discourse concerning transhumanism, who wants to "enhance" humans technologically. However, "enhancement" is usually meant in a masculine and capitalist regime of enhancement: think faster, jump higher, live longer, be more productive.

This chapter makes the claim that present debates around AI and transhumanism are characterized by a general lack of sensitivity and critical distance to the various levels and contexts of metaphors used. As a consequence, questions regarding responsibility and potential use cases are greatly distorted. The chapter proves its claim by critically analysing core notions of the AI and transhumanism debates and is a conceptual work at the intersection of computer science/machine learning and the philosophy of language.

R. Rehak (✉)
Weizenbaum Institute for the Networked Society, Berlin, Germany
e-mail: rainer.rehak@wzb.eu

W. Hofkirchner, H.-J. Kreowski (eds.), *Transhumanism: The Proper Guide to a Posthuman Condition or a Dangerous Idea?*, Cognitive Technologies,
https://doi.org/10.1007/978-3-030-56546-6_11

1 Introduction

In the 1970s, the British physicist and science fiction writer Arthur C. Clarke coined the phrase of any sufficiently advanced technology being indistinguishable from magic—understood here as mystical forces not accessible to reason or science. In his stories, Clarke often described technical artefacts such as anti-gravity engines, "flowing" roads or tiny atom-constructing machinery. In some of his stories, nobody knows any more how exactly those technical objects work or how they have been constructed; they just use them and are happy doing so.

In today's specialized society with a division of labour, most people also do not understand most of the technology they use. However, this is not a serious problem, since for each technology there are specialists who understand, analyse and improve the products in their field of work—unlike in Clarke's worlds. But since they are experts in few areas and people's lifetime is limited, they are, of course, laypersons or maybe hobbyists in all other areas of technology.

After the first operational universal programmable digital computer—the Z3—had been invented and built in 1941 in Berlin by Konrad Zuse, the rise of the digital computer into today's omnipresence started. In the 1960s, banks, insurances and large administrations began to use computers; police and intelligence agencies followed in the 1970s. Personal computers appeared and around that time newspapers wrote about the upcoming "electronic revolution" in publishing. In the 1980s, professional text work started to become digital, and in the 1990s, the Internet was opened to the general public and to commercialization. The phone system became digital, mobile Internet became available and, in the mid-2000s, smartphones started to spread across the globe (Passig and Scholz 2015).

During the advent of computers, they were solely operated by experts and used for specialized tasks such as batch calculations and book-keeping on a large scale. Becoming smaller, cheaper, easier to use and more powerful over time, more and more use cases emerged up to the present situation of computer ubiquity. More applications, however, also entail more impact in personal lives, commercial activities or even societal change. The broader and deeper the effects of widespread use of networked digital computers became, the more pressing political decisions about their development and regulation became as well.

The situation today is characterized by non-experts constantly using computers, sometimes not even noticing it, and non-experts making decisions about computer use in business, society and politics, from schools to solar power, from cryptography to cars. The only way to discuss highly complex computer systems and their implications is by analogies, simplifications and metaphors. However, condensing complex topics into understandable, discussable and then decidable bits is difficult in at least two ways. First, one has to deeply understand the subject and, second, one has to understand its role and context in the discussion to focus on the relevant aspects (Coy 1992). The first difficulty has to do with knowledge and lies in the classical technical expertise of specialists. But the second difficulty concerns what exactly should be explained in what way. Depending on the context of the discussion, certain aspects of the matter have to be explicated using explanations, metaphors and analogies highlighting the relevant technical characteristics and implications. Seen in this light, this problem of metaphors for technology is not only philosophically highly interesting but also politically very relevant. Information technology systems are not used because of their actual technical properties, but because of their assumed functionality, whereas the discussion about the functionality is usually part of the political discourse itself. Given the complexity of current technology, only experts can understand such systems, yet, only a small number of them actively and publicly take part in corrective political exchanges about technology. Especially in the fields of artificial intelligence (AI) and transhumanism a wild jungle of metaphors is in use. However, as long as discussions take place among specialists treating the metaphors as domain-specific technical vocabulary, no harm is done. But domain-specific language often diffuses into other fields.

2 Conceptual Domains and Everyday Language

Unlike the abstract field of mathematics, where most technical terms are easily spotted as such, AI makes heavy use of anthropomorphisms. Considering AI terms such as "recognition", "learning", "acting", "deciding", "remembering", "understanding" or even "intelligence" itself, problems clearly loom all across possible conversations. Of course, many other sciences—especially the social sciences—also use scientific terms that are derived from everyday language. In this case, these words then have clearly defined meanings or at least linked discourses reflecting upon them. Examples are the terms "fear" in psychology, "impulse" in physics, "will" in philosophy or "rejection" in geology and "ideology" in mathematics. Often the same words have completely different meanings in different domains, sometimes even contradictory meanings, as the examples of "work" in physics and economic theory or "transparency" in computer science and political science illustrate.

Hence, problems arise when these scientific terms are transferred carelessly into other domains or back into everyday language used in political or public debates (Bonsiepen 1994). This can occur through unprofessional science journalism, deliberate inaccuracy for PR purposes, exaggerations for raising third-party funding or generally due to a lack of sensitivity to the various levels and contexts of metaphors (Rehak 2019).

3 The Case of Artificial Intelligence

For some years now, technical solutions utilizing artificial intelligence are widely seen as a means to tackle many fundamental problems of mankind—from fighting the climate crisis, tackling the problems of ageing societies, reducing global poverty, stopping terror, detecting copyright infringements or curing cancer to improving evidence-based politics, improving predictive police work, local transportation, self-driving cars and even waste removal (Rehak 2018). Although there is no narrow definition of artificial intelligence, the range of assigned functionality reaches from applying traditional statistics to using machine learning (ML) techniques or even generally to "highly complex information systems", as in the official "Social Principles of Human-centric AI" of Japan (Council for Social Principles of Human-centric AI 2019). In the following, we will concentrate on a specific subclass of machine learning techniques—artificial neural networks (ANN)—to illustrate the fallacies and pitfalls of questionably used metaphors. Of course, the problems mentioned also apply to other forms of AI, when a similar terminology is being used.

Key drivers for the current AI renaissance are the successes of applying artificial neural networks to huge amounts of data now being available and using new powerful hardware. Although the theoretical foundations of the concepts used were conceived as early as the 1980s, the performance of such systems has improved to such an extent over the last few years that they can now be put to practical use in many new use cases, sometimes even in real-time applications such as image or speech recognition. Especially if huge datasets for training are available, results can be much better than traditional symbolic approaches.

Before we analyse the language being used to describe the functionality, we should have a look at the inner workings of artificial neural networks to have a base for scrutinizing terminologies.

4 Basic Structure of Artificial Neural Networks

Artificial neural networks are an approach of computer science to solve complex problems that are hard to explicitly formulate or, more concretely, to program. Those networks are inspired by the function of the human brain and its network of neurons; however, the model of a neuron being used is very simplistic. Many details of biological neuronal networks, such as myelinization or ageing, are left out. Following the original model, each artificial neuron, the smallest unit of such systems, has several inputs and one output. In each artificial neuron, the inputs are weighted according to its configuration and then summed up. If the result exceeds a certain defined threshold the neuron is triggered, and a signal is passed on to the output. These neurons are usually formed into "layers", where each layer's outputs are the next layer's inputs. The resulting artificial neural network thus has as its input the individual inputs of the first layer and as its output the individual outputs of the last layer. The layers in between are usually called "hidden" layers and with many hidden layers an artificial neural network is usually called "deep".

In the practical example of image recognition, the input would consist of the colour values of all distinct pixels in a given image and the output would be the probability distribution among the predefined sets of objects to recognize.

5 Configuring the Networks

There are various ways of configuring artificial neural networks, which will now briefly be described. Building such a network involves certain degrees of freedom and hence decisions, such as the number of artificial neurons, the number of layers, the number and weights of connections between artificial neurons and the specific function determining the trigger behaviour of each artificial neuron. To properly recognize certain patterns in the given data—objects, clusters, etc.—all those parameters need to be adjusted to a use case. Usually, there are best practices how to initially set it up; then the artificial network has to be further improved step by step. During this process, the weights of the connections will be adjusted slightly in each step, until the desired outcome is created, may it be the satisfactory detection of cats in pictures or the clustering of vast data in a useful way. Those training cycles are often done with a lot of labelled data and then repeated until the weights do not change any more. Now it is a configured artificial neural network for the given task in the given domain.

6 Speaking About the Networks

Now we will take a closer look how computer scientists speak about this technology in public and how those utterances are carried into journalism and furthermore into politics. As mentioned above, the description of artificial neutral networks as being inspired by the human brain already implies an analogy which must be critically reflected upon. Commonly used ANNs are usually comparatively simple, both in terms of how the biochemical properties of neurons are modelled and in the complexity of the networks themselves. A comparison: The human brain consists of some 100 billion neurons, while each is connected to 7000 other neurons on average. ANNs on the other hand are in the magnitude of hundreds or thousands of neurons, while each is connected to tens or hundreds of other neurons. This difference in orders of magnitude entails a huge difference in functionality, let alone misunderstanding them as models of the human brain. Even if to this point the difference might only be a matter of scale and complexity, not one of principle, we have no indication of that changing anytime soon. Thus, using the notion of "human cognition" to describe ANN is not only radically oversimplifying, it also opens up the metaphor space to other neighbouring yet misleading concepts. For example, scientists usually do not speak of networks being configured but being "trained" or doing "(deep) learning". Along those lines are notions like "recognition", "acting", "discrimination", "communication", "memory", "understanding" and, of course, "intelligence".

7 Considering Human Concepts

When we usually speak of "learning", it is being used as a cognitive and social concept describing humans (or, to be inclusive, intelligent species in general) gaining knowledge with an individual learner or a group, other peers, motivations, intentions, teachers or coaches, the context and a whole range of learning methods being researched, tested and applied in the academic and practical fields of psychology of learning, pedagogy or educational science (Piaget 1944). This is substantially different to the manual or automated configuration of an ANN using test sets of data. Seen in this light, the common notion of "self-learning systems" sounds even more misplaced. This difference in understanding has great implications, since e.g. an ANN would never get bored with its training data and therefore decide to learn something else or simply refuse to continue; metaphors matter, no Terminator from the movies in sight.

"Recognition" and "memory" are also very complex concepts in the human realm. Recognizing objects or faces requires attention, focus, context and—depending on one's school of thought—even consciousness or emotions. Human recognition is therefore completely different from automatically finding differences of brightness in pictures to determine the shape and class of an expected object (Rispens 2005). Further, consciously remembering something is a highly complex process for humans which is more comparable to living through imagined events again and by that even changing what is being remembered. Human memory is therefore a very lively and dynamic process, and not at all comparable to retrieving accurate copies of stored data bits.

Especially the notions of "action" or even "agency" are highly problematic when being applied to computers or robots. The move of a computer-controlled robotic arm in a factory should not be called a robot's "action", just because it would be an "action" if the arm belonged to a human being. Concerning human actions, very broad and long-lasting discussions at least in philosophy and the social sciences already exist. Note the difference between "behaviour" and "action": The former only focuses on observable movement, whereas the latter also includes questions of intention, meaning, consciousness, teleology, world modelling, emotions, context, culture and much more. While a robot or a robotic arm can be described in terms of behavioural observations, its movements should not easily be called actions.

Similarly, complex is the notion of "communication" in a human context, since communication surely differs from simply uttering sounds or writing shapes. "Communication" requires a communication partner, who knows that the symbols used have been chosen explicitly knowing that they will be interpreted as deliberate utterances (Goodman 1976). Communication therefore needs a common understanding of the communicational situation by the involved parties. Hence, the sound of a loudspeaker or the text on a screen does not constitute a process of communication in the human sense, even if their consequences are a transfer of information. If there is no reflection of the communication partner, no deliberation, no freedom as to which symbols to choose and what to communicate, one should not easily apply such complex notions as "communication" outside its scope without explanation (Von Foerster 1995).

Furthermore, the concept of "autonomy"—as opposed to heteronomy or being externally controlled—is widely used nowadays when dealing with artificial intelligence, may it be concerning "intelligent" cars or "autonomous" weapon systems. Although even human autonomy has been largely criticized, some even say completely deconstructed, in the social sciences in the last century since individuals are largely influenced by culture, societal norms and the like, autonomy seems to gain new traction in the context of computer science. Yet, it is a very simplistic understanding of the original concept (Kreowski 2018). Systems claimed to be "autonomous" heavily depend on many factors, e.g. a stable, calculable environment, but also on programming, tuning, training, repairing, refuelling and debugging, which are still traditionally done by humans, often with the help of other technical systems. In effect, those systems act according to inputs and surroundings, but they do not "decide" on something. Here again, the system can in principle not contemplate its actions and finally reach the conclusion to stop operating or change its programmed objectives autonomously. Hence, artificial intelligence systems—with or without ANN—might be highly complex systems, but responsibility or accountability should not be attributed to them. Here we see one concrete instance of the importance of differentiating between narrow (domain-specific) AI and general (universal) AI. This clarification is not meant to diminish the technical work of all engineers involved in such "autonomous" systems; it is purely a critique about how to publicly talk about such systems and its capabilities in non-expert contexts.

8 Instances and Consequences

After having briefly touched upon some areas of wrongly used concepts, we can take a look at a concrete example, where such language use specifically matters.

A very interesting and at that time widely discussed example was Google's "Deep Dream" image recognition and classification software from 2015, code name "Inception" (Mordvintsev and Tyka 2015). As described above, artificial neural networks do not contain any kind of explicit models; they implicitly have the "trained" properties distributed within their structures. Some of those structures can be visualized by inserting random data—called "noise"—instead of actual pictures. In this noise, the ANN then detects patterns exposing its own inner structure. Interesting for this paper are not the results—predominantly psychedelic imagery—but the terminology being used in Google's descriptions and journalists' reports. The name "Deep Dream" alone is already significant, but also the descriptive phrases "Inside an artificial brain" and "Inceptionism". Both (deliberately) give free rein to one's imagination. In additional texts provided by Google, wordings such as "the network makes decisions" accumulate. Further claims are that it "searches" for the right qualities in pictures, it "learns like a child" or it even "interprets" pictures. Using this misleading vocabulary to describe ANNs and similar technical artefacts, one can easily start to hope being able to learn something about the fundamentals of human thinking. Maybe those texts and descriptions have been written for the primary purpose of marketing or public relations, since they explain little but signify the abilities and knowledge of the makers, yet that does not diminish the effect of the language used. For many journalists and executive summary writers or even the interested public those texts are the main source of information, not the actual scientific papers. In effect, many of those misleading terms were widely used, expanded on and by that spread right into politician's daily briefings, think tank working papers and dozens of management magazines, where the readers are usually not aware of the initial meanings. This distorted "knowledge" then becomes the basis for impactful political, societal and managerial decisions.

Other instances where using wrong concepts and wordings mattered greatly are car crashes involving automated vehicles, e.g. from companies like Uber, Google or Tesla. For example in 2018, a Tesla vehicle drove into a parked police car in California, because the driver had activated the "autopilot" feature and did not pay attention to the road any more. This crash severely exposed the misnomer. The driver could have read the detailed "autopilot" manual before invoking such a potentially dangerous feature, yet, if this mode of driving would have been called "assisted driving" instead of "autopilot", very few people would have expected the car to autonomously drive "by itself". So thinking about a car having an autopilot is quite different from thinking about a car having a functionality its makers call "autopilot". Actually, reading into Tesla's manuals, different levels of driving assistance are being worked on, e.g. "Enhanced Autopilot" or "Full Self-Driving", whereas the latter has not been implemented so far. Further dissecting the existing "autopilot" feature, one finds it comprises different sub-functionalities such as Lane Assist, Collision Avoidance Assist, Speed Assist, Auto High Beam, Traffic Aware Cruise Control or Assisted Lane Changes. This collection of assistance technologies sounds very helpful, yet it does not seem to add up to the proclaimed new level of autonomous driving systems with an autopilot being able to "independently" drive by itself.

Those examples clearly show how a distinct reality is created by talking about technology in certain terms yet avoiding others. Choosing the right terms is not always a matter of life and death, but they certainly pre-structure social and societal negotiations regarding the use of technology (Bonsiepen 1994).

9 Malicious Metaphors

Suddenly we arrive in a situation where metaphors are not only better or worse for explaining specifics of technology, but where specific metaphors are deliberately being used to push certain agendas, in Tesla's case to push a commercial and futurist agenda—commercial because of using "autonomy" as unique selling point and futuristic as it implies that "autonomy" is a necessary and objective improvement for everyone's life and society as a whole. Generally most innovative products involving "artificial intelligence" and "next-generation technology" are being communicated as making "the world a better place", "humans more empowered" or "societies more free".

10 Transhumanism and Regimes of Enhancement

So the next time decision-makers and journalists will be asked about possibilities of technology, they will surely remember having heard and read about computers winning Chess and Go, driving cars, recognizing speech, translating text, managing traffic and generally finding optimal solutions to given problems (Dreyfus 1972). But using deficient anthropomorphisms like "self-learning", "autonomous" or "intelligent" to describe the technical options of solving problems is only one part of the misleading metaphor problem. Sadly, the notional sloppiness goes both ways: Taking a closer look at debates within the tech sector when dealing with the societal use of technology, the (non-technical) goals themselves tend to often be analysed and described using a mix of technological and solutionist terms, as people like R. Kurzweil, E. Musk and others influentially do. This is especially true for the transhumanist debate.

Usually the technical criteria for optimally reaching a goal can be easily explicated: faster processing, less memory usage, longer durability, better scalability, more precision or higher energy efficiency. But how about finding and defining the goal itself? To what ends should those means be used and which terminology is used to formulate those ends? When we pose the transhumanist question regarding how information technology can help human beings the answer is usually "enhancement". Yet, the notion of "enhancement" is being used in a very technical way, ignoring its fundamental ambiguity. With information technology, so the argument from the classic flavour of transhumanism goes, we will soon be able to "fix and update the human operating system" (Kurzweil 2005): make our brain remember more faces, forget less details, think faster, jump higher, live longer, see more sharply, be awake longer, be stronger, hear more frequencies and even create new senses—exactly how a technologist would imagine what new technology could deliver for humanity (Coy 1993).

The underlying assumption is a very specific—to be precise technical—understanding of what is considered "good" or "desirable". But does every human or even the majority primarily want to remember more, forget less, live longer or run faster? Are those aspects even the most pressing issues we want technology to solve? In addition, not only do those fantasies happily follow along the lines of the neo-liberal logic of applying quantification, competition, performance and efficiency into all aspects of life, they also unconsciously mix in masculinist—even militarist—fantasies of power, control, strength and subjugation of the natural or finally correcting the assumed defective.

As valid as those opinions concerning optimizations are, still it is important to note that views like that imply absolute values and are incompatible with views which put social negotiation, non-mechanistic cultural dynamics or in general pluralistic approaches in their centre. I call those conflicting groups of views *regimes of enhancement*. It is clearly not possible to "enhance" a human being with technically actualized immortality, if one does not want to live forever or does not find it particularly relevant. Many other conflicting views can be thought of. However, the mere acceptance of the concept of *regimes* already breaks any claim for absoluteness and opens the door for discussing different understandings of "enhancements". Accepting this already makes positions somehow compatible and allows for individual or even societal endeavours of creative re-interpretations of the concept of transhumanism itself.

To summarize, as sloppy as the anthropomorphic metaphors for describing technical methods are as blurry also the actual individual or societal problems to solve are being described in (unfitting) technical terms. Taking both shifts seriously helps to understand why so many problems only seem to have an easy technical solution (Morozov 2013).

11 Closing Remarks

Technology is used and politically decided upon perceived functionality, not upon the actually implemented functionality. However, communicating functionality is much more driven by interests than creating the actual technology already is. Therefore, attribution ascription is a very delicate and consequential issue that paints a differentiated picture of the consequences of careless use of terms. If relevant decision-makers in politics and society are (really) convinced at some point that these "new" artificial neural networks can develop an understanding of things or properly interpret facts, nothing would stand in the way of their use for socially or politically sensitive tasks like deciding about social benefits or judging court cases. Here the difference between metaphorical "judging" and real judging, metaphorical "acting" and real acting plays out (Weizenbaum 1976). If one acts in the social science meaning of the word, one has to take responsibility for one's actions, if a computer only "acts", used as a metaphor, responsibility is blurred.

Hence, especially scientific journalists but also computer professionals have the professional responsibility to be more sensitive about proper imagery being used, and misleading metaphors being criticized. The danger here does not lie in consciously not understanding computers or AI but in not understanding them while thinking having understood them. Finally, a chess computer will never get up and change its profession to play Go, exponential growth in computing power does so far not entail more than slight growth of "cognitive" functionality and the fear of computers eliminating all human jobs is a myth capable of inciting fear since at least 1972 (Dreyfus 1972; Butollo 2018).

If technological discussions and societal reflections on the use of technology are to be fruitful, scientists and journalists alike have to stop joining the "buzzword-driven language game", which does neither help with solutions nor does it advance science. It merely entertains our wishful thinking how magical technology should shape the future. Indeed, any sufficiently advanced technology is indistinguishable from magic—to the layperson—, but we also have to conclude that this "magic" is being constructed and used by certain expert "magicians" to advance their own interests and agenda or that of their masters. So not even such magic spares us the necessity to pay attention to power, details and debate.

References

Bonsiepen, L.: Folgen des Marginalen. In: Die maschinelle Kunst des Denkens – Perspektiven und Grenzen der KI. Vieweg, Braunschweig/Wiesbaden (1994)

Butollo, F.: Automatisierungsdividende und gesellschaftliche Teilhabe, Essay, Erschienen auf: regierungsforschung.de. NRW School of Governance, Duisburg (2018)

Council for Social Principles of Human-centric AI. Social principles of human-centric AI, Japan (2019)

Coy, W.: Für eine Theorie der Informatik! In: Sichtweisen der Informatik. Vieweg, Braunschweig/Wiesbaden (1992)

Coy, W.: Reduziertes Denken. Informatik in der Tradition des formalistischen Forschungsprogramms. In: Informatik und Philosophie. Mannheim, BI Wissenschaftsverlag (1993)

Dreyfus, H.L.: What Computers Can't Do. Harper & Row, New York (1972)

Goodman, N.: Languages of Art: An Approach to a Theory of Symbols. Hackett Publishing, London (1976)

Kreowski, H.-J. Autonomie in technischen Systemen, Leibniz Online, Nr. 32, Zeitschrift der Leibniz-Sozietät e. V (2018)

Kurzweil, R.: The Singularity Is Near: When Humans Transcend Biology. Penguin, London (2005)

Mordvintsev, A., Tyka, M.: Inceptionism: going deeper into neural networks. https://ai.googleblog.com, https://ai.googleblog.com/2015/06/inceptionism-going-deeper-into-neural.html (2015). Accessed 17 June 2020

Morozov, E.: To Save Everything, Click Here: The Folly of Technological Solutionism. Penguin, London (2013)

Passig, K., Scholz, A.: Schlamm und Brei und Bits – Warum es die Digitalisierung nicht gibt. Klett-Cotta Verlag, Stuttgart (2015)

Piaget, J.: Die geistige Entwicklung des Kindes. M.S. Metz, Zürich (1944)

Rehak, R.: Technische Lösung sucht Problem. Zukunftsfähige Infrastrukturen im städtischen Bereich. In: Politische Ökologie, vol. 155. oekom verlag, München (2018)

Rehak, R.: A Trustless society? A Political look at the blockchain vision. In: Beiträge zur Hochschulforschung, vol. 41, pp. 60–65. Bayerisches Staatsinstitut für Hochschulforschung und Hochschulplanung, München (2019)

Rispens, S. I. Machine Reason: A History of Clocks, Computers and Consciousness (2005)

Weizenbaum, J.: Computer Power and Human Reason: from Judgment to Calculation. W. H. Freeman, San Francisco (1976)

Von Foerster, H.: Cybernetics of Cybernetics, the Control of Control and the Communication of Communication. Carl Auer Systeme Verlag, New York (1995)

Part IV
Sociological Aspects

Chapter 12
Transhumanism and/as Whiteness

Syed Mustafa Ali

Abstract Transhumanism is interrogated from critical race theoretical and decolonial perspectives with a view to establishing its "algorithmic" relationship to historical processes of race formation (or racialization) within Euro-American historical experience. Although the transhumanist project is overdetermined vis-à-vis its raison d'être, it is argued that a useful way of thinking about this project is in terms of its relationship to the shifting phenomenon of whiteness. It is suggested that transhumanism constitutes a techno-scientific response to the phenomenon of "White Crisis" at least partly prompted by contestation of Eurocentrically universal humanism.

1 Introduction

In a widely cited poststructuralist/anti-humanist critique of European humanism,[1] Badmington (2003) argues that "there is nothing more terrifying than a posthumanism that claims to be terminating 'Man' while actually extending 'his' term in office" (p. 16). In this prescient statement, attention is drawn to the very real possibility of a posthumanist orientation that, while claiming to be *critical*, ends up re-inscribing precisely that very humanism—focused on the figure of "Man" as white, male, European and anthropocentric—that it sets out to challenge ("post-" *as* dialectical engagement) and overcome ("post-" *as* temporal/historical

[1]Badmington's argument is informed by various critical currents within contemporary European thought including postcolonial theory and a commitment to the post-discursive, "new materialist" embrace of boundary-disrupting ontological affinity with the non-human (animal, machine, etc.) associated with the "ontological turn". It is also motivated by a concern to address the political and ecological implications of the anthropocentrism and subject–object dualism associated with dominant strands of Enlightenment thought.

S. M. Ali (✉)
School of Computing and Communications, The Open University (UK), Milton Keynes, UK
e-mail: s.m.ali@open.ac.uk

W. Hofkirchner, H.-J. Kreowski (eds.), *Transhumanism: The Proper Guide to a Posthuman Condition or a Dangerous Idea?*, Cognitive Technologies,
https://doi.org/10.1007/978-3-030-56546-6_12

transcendence to a new ontological condition). In what follows, I attempt to think through some possible implications of Badmington's claim by exploring the entangled relationship between techno-scientific conceptions of the posthuman, transhumanism and the phenomenon of whiteness against the background of what is, ostensibly, a contemporary resurfacing—or *re*-iteration—of the historical phenomenon of "White Crisis" with the aim of mounting a decolonial critique[2] of the transhumanist/techno-scientific posthumanist project.

Badmington (2003) has argued that "apocalyptic accounts of the end of 'Man' . . . ignore humanism's capacity for regeneration and, quite literally, recapitulation" (p. 11). Against this, I want to suggest that it is the very apocalyptic[3] nature of the phenomenon of White Crisis—that is, perceived threat to white supremacy under mounting contestation from the non-white "other"—that contributes to[4] engendering what I refer to as the "algorithmic" transformation of humanism into posthumanism via transhumanism as an *iterative shift* within the historically sedimented onto-logic of Eurocentric racialization. My point of departure turns on the "between-ness" of the transhuman[5] vis-à-vis the posthuman, such that the former is engaged against the background provided by the latter as *telos*, irrespective of how this is ultimately realized in techno-scientific terms, viz. augmented biological form, uploaded mind or synthetic, artificial intelligence—that is, "Mind Children". Engaging transhumanism as an *iteration* within the algorithmic logic(s) of race/racism/

[2]By a "decolonial" critique, I mean one that foregrounds considerations of the body-politics (who) and geo-politics (where) of knowing and being and is preferentially disposed towards thinking through conceptual frameworks emerging from the periphery (margins, borders) of the modern/colonial world system.

[3]There is a secular Enlightenment rationalist tendency to dismiss apocalyptic narratives as an irrational hangover from "the age of religion"; however, as Gray (2007) convincingly argues, apocalyptic and utopian thinking derived from the Christian tradition informs secular frameworks, both those on the conservative "right" and those on the critical "left". In addition, there is the need to consider Noble's (1997) and Davis' (1998) exploration of the long durée "entanglement" of apocalyptic religious and occultist thinking with scientific and technological development in the European/"Western" tradition. For a preliminary exploration of the entangled nature of Western apocalypticism, race, religion, transhumanism and AI, see (Ali 2019).

[4]By framing the issue in terms of *contribution* rather than *causation*, I recognize that the transhumanist/technological posthumanist project is overdetermined in terms of its historical motivations and causes. Insofar as ideas of leveraging technology to achieve utopian and/or apocalyptic purposes have a long history, I am *not* suggesting that the transhumanist project is driven *solely* by a post-racial crisis of whiteness; rather, I argue that under contemporary conditions of White Crisis, the transhumanist project gains a sense of urgency as a techno-scientific resolution—or "fix"—to such an anxiety-ridden state of affairs.

[5]According to Bostrom (2014), "in its contemporary usage, 'transhuman' refers to an intermediary form between the human and the posthuman" (p. 4). Transhumanism is generally framed in terms of the application of GRIN (genetics, robotics, information technology and nanotechnology) in the service of self-directed evolution—that is, enhancement of the human—towards a technocratic future. A related acronym is NBICS which refers to the combined resources of nanotechnology, biotechnology, information technology, cognitive science and synthetic biology.

racialization associated with colonial modernity,[6] I maintain that the emergence of the techno-scientific posthuman points to a transformation in the nature of humanism that maintains structurally asymmetric power relations between the (formerly) human (as white, Western, male, etc.) and the subaltern, sub-human "other" even as the latter contests the Eurocentrism of "the human".[7]

Following a discussion of the methodological precedents informing this study, I go on to briefly explore the nature and genealogy of transhumanism, the demographic constitution of the transhumanist movement and the meaning of whiteness and White Crisis with a view to situating transhumanism and technological posthumanism as developments within the logic(s) of algorithmic racism.

2 Methodological Precedents

In the context of exploring race "and/as" technology, Chun (2009) maintains that race *as* technology "shifts the focus from the *what* of race to the *how* of race, from *knowing* race to *doing* race by emphasizing the similarities between race and technology"; further, that "*race as technology* is a simile that posits a comparative equality or substitutability—but not identity—between the two terms" (p. 8). Drawing inspiration from Chun's engagement with race and/as technology, and building on earlier work reflexively exploring other related "as/and" configurations such as race and/as information (Ali 2013) and Orientalism and/as information (2015), informed by a critical race theory of information (Ali 2012) and decolonial computing perspective (Ali 2014, 2016a) which motivate consideration of the entanglement of race, religion, information, computing and related ICT phenomena with the body-politics and geo-politics (and theo-politics) of knowing and being, I aim to critically interrogate transhumanism as a techno-scientific response to the phenomenon of White Crisis at least partly prompted by contestation of Eurocentrically universal humanism.

[6]By "colonial modernity" is meant the condition associated with the world system emerging during the long durée of the sixteenth century CE. From a decolonial and critical race theoretical perspective, this system must be understood as *both* modern *and* colonial insofar as European colonialism played a *constitutive* role in the emergence of the modern world and is its "dark underside" (Mignolo 2011). In addition, it is crucial to appreciate that *coloniality*—that is, the facilitating structuring logics (ontological, epistemological, cultural, political, economic, etc.) of the colonial project—persists in the postcolonial era notwithstanding the formal end of colonialism with the national independence movements of the 1960s.

[7]As Badmington (2003) states, "the seemingly posthumanist desire to download consciousness into a gleaming digital environment is itself downloaded from the distinctly humanist matrix of Cartesian dualism. Humanism survives the apparent apocalypse and, more worryingly, fools many into thinking that it has perished. Rumours of its death are greatly exaggerated" (p. 11); in short, "the new *now* secretes the old *then*. Humanism remains" (p. 14). In this connection, and somewhat anticipating the argument presented herein, Islam (2014) poignantly remarks that "today's subaltern is tomorrow's human or pre-posthuman" (p. 5).

While Chun's concern is to posit a comparative equality or substitutability between race and technology, my focus lies in exploring the implications of positing a similar "comparative equality or substitutability" between two different yet related terms, however, one in which the ordering of terms is inverted somewhat vis-à-vis the ordering suggested by Chun, viz. transhumanism and/as whiteness, thereby engaging the issue of how transhumanism might be thought about in relation to processes of racialization—specifically, those associated with the largely tacit background phenomenon of a hegemonic whiteness. Chun maintains that "by framing questions of race *and* technology, as well as by reframing race *as* technology, in relation to modes of media naturalization [we can] theoretically and historically better understand the force of race and technology and their relation to racism" (p. 8). Arguing along similar lines, I want to suggest that framing questions of transhumanism *and* whiteness, as well as reframing transhumanism *as* whiteness, in relation to historical processes of re-articulation of the latter (that is, whiteness), enables us to theoretically and better understand how transhumanism can—and arguably *does*—function as a techno-scientific articulation of whiteness during a period arguably marked by increasing contestation of other forms of this racial phenomenon.[8]

[8]Somewhat optimistically, Coleman (2009) has argued that "technology's embedded function of self-extension may be exploited to liberate race from an inherited position of abjection toward a greater expression of agency" (p. 177); on her view, by "extending the function of *techné* to race, I create a collision of value systems. In this formulation, race exists as if it were on par with a hammer or a mechanical instrument; *denaturing it from its historical roots, race can then be freely engaged as a productive tool*. For the moment, let us call 'race as technology' a disruptive technology that changes the terms of engagement with an all-too-familiar system of representation and power [emphasis added]" (p. 178). Notwithstanding their possible rhetorical value vis-à-vis engaging in projects of decolonially and critical race theoretically informed resistance to systemic/structural racism—more specifically, global white supremacy—I want to suggest that such assertions are problematic on account of an ostensibly tacit assumption that technology stands separate from, rather than entangled with, race. What appears to be missing from Coleman's (and Chun's) formulation is reflexive consideration of technology and/as race—that is, recognition of the *racialized* ontology of technology under colonial modernity, and I suggest that this follows directly from the "bracketing" of the historical that Coleman is committed to embracing in her "technological turn". Yet my critique of their position should not be understood as entailing support for the view that technology is *necessarily*, in the sense of *trans*-historically, racialized since that line of argument turns on the questionable assertion of the trans-historicity of race itself; on the contrary, a commitment to the *contingency* of technology's racialization is maintained, yet one that requires us to consider more seriously how the field of technology/technique is racially inflected, such racial inflection contributing to a *historical* essence that in colonial modernity has a racialized underside, and which thereby constraints/limits scope for resistant action vis-à-vis affording non-abject possibilities for racial agency.

3 Transhumanism and Its Genealogy

Given the vast and ever-expanding discursive terrain associated with transhumanism, in what follows I confine myself to briefly expounding some representative formulations of the concept—specifically, those which shed light on its genealogy, and which are framed in terms of transformative shifts about the figure of the human. Common to all such conceptions is the relationship between transhumanism, the European Renaissance and Enlightenment humanism. According to Bostrom (2011), transhumanism is a *continuation* of eighteenth-century European Enlightenment commitments—specifically, scientific empiricism and rational humanism, and in a later work, he argues that "transhumanism can be viewed as an *extension* of humanism, from which it is partially derived [emphasis added]" (Bostrom 2014, p. 1). This position is supported by Jotterand (2010), Hughes (2012), Ferrando (2013) and others, the latter of whom maintains that "emphasis on notions such as rationality, progress and optimism is in line with the fact that, philosophically, transhumanism roots itself in the Enlightenment" (p. 27). Consider, in this connection, extropian theorist Max Moore's (2013) invitation to think about transhumanism as "trans-humanism" plus "transhuman-ism". According to this view,

> 'Trans-humanism' emphasizes the philosophy's roots in Enlightenment humanism. From here comes the emphasis on progress (its possibility and desirability, not its inevitability), on taking personal charge of creating better futures rather than hoping or praying for them to be brought about by supernatural forces, on reason, technology, scientific method, and human creativity rather than faith ... 'Trans-human' emphasizes the way transhumanism goes well beyond humanism in both means and ends. Humanism tends to rely exclusively on educational and cultural refinement to improve human nature whereas transhumanists want to apply technology to overcome limits imposed by our biological and genetic heritage. (p. 4)

Bostrom (2014) complements this perspective by maintaining that "[the] 'transhuman' refers to an *intermediary* form between the human and the posthuman [emphasis added]" (p. 4), whereby the latter is meant "possible future beings whose basic capacities so radically exceed those of present humans as to be no longer unambiguously human by our current standards" (p. 3). This conceptualization of transhumanism as a transitional precursor to posthumanism is supported by Sombetzki (2016) who maintains that "Transhumanism can be seen as a *link* between traditional humanism and posthumanism [insofar as] Transhumanists still hold on to the principle of individual perfection and rational meliorism as humanists do" (p. 171). However, she insists that it is a mistake—more precisely, a "distortion" or "perversion" of humanism—to argue for a relationship of continuity (or extension) between Enlightenment humanism and technological posthumanism, wherein the latter is seen as driven by the convergence of GRIN/NBICS technologies and culminates in mind uploading.[9] According to Sombetski, Enlightenment

[9]In this connection, Sombetzki (2016) maintains that technological posthumanism is a "perverted humanism" because "it perverts the idea of individual perfection and rational meliorism in stating

humanism was predicated on three commitments, viz. (1) mind–matter dualism, (2) autonomous human rationality and (3) mortality of matter/the body, and fundamentally concerned with self-cultivation of the intellect or the rational nature of man (p. 168).[10] However, all three commitments are arguably subjected to contestation by proponents of technological posthumanism leading Sombetski to assert that

> [Technological] posthumanists aren't transhumanists in [the] sense that they are not interested in enhancing or perfecting man as he currently exists. [Technological] posthumanists' goal is a new race—whether mechanical or organic. But commonly, they speak in the first person plural when it comes to mind uploading: 'we' will be uploaded one day, 'we' will get online and exist as a virtual unity when the era of the Singularity finally dawns on the morning of the universe. It is this 'we' that reveals [technological] posthumanism as a *perverted humanism*. [Technological] posthumanists want the new breed, but they want 'us' to still be there individually, they *want* 'us' as individuals and want 'us' *not* to be that new race. (p. 173)

While conceding the importance of drawing attention to the emergence of an "us–them" distinction between humans/transhumans ("us") and technological posthumans ("them"), it is somewhat ironic to note that Sombetzki fails to engage with a prior, and arguably *grounding*, "us–them" distinction operative within the human group with which posthumans contrast themselves, viz. that between those humans classified as the non-European/non-Western/non-white "other"/"them" against and in terms of which the European/Western/white "self"/"us" historically was—and arguably still is—relationally constituted through a process of antagonistic negative dialectics.[11] In addition to this oversight, which turns on a failure to consider Enlightenment modernity in relation to its "dark underside" of racial colonialism (Mignolo 2011), Sombetski fails to adequately engage with modernity's *other* "dark underside", viz. that pertaining to the mythic and religious (Mahootian 2012). While drawing attention to "mythological and ancient elements that most obviously merge in the enlightenment concept of human nature" (p. 164), specifically the *Epic of Gilgamesh*,[12] Sombetski goes on to refer to the ideas of Renaissance

the ultrasupremacy of a genuinely immortal human mind as genuinely independent from its biological heritage" (p. 174).

[10]While it is beyond the scope of this essay to explore this issue at length, it is important to appreciate that a certain Eurocentric conflation of terms—specifically, rationality and intellect—is operative within Sombetski's discourse. For a useful critique of the tendency to conflate human intellect with Enlightenment reason/rationality, see Ogunnaike (2016).

[11]In this connection, critical philosophers of race maintain that it is against such an "other"/"them" posited as a *group/collective* phenomenon that the *individuality* of rational autonomous humans (European, white, Western) is conceived.

[12]Appeal to this Babylonian myth—and perhaps from a critical race theoretical and/or decolonial perspective this might be better understood as an instance of *appropriation*—is common among proponents of transhumanism. In this connection, Hughes (2012) is led to assert that "Transhumanism is a modern expression of ancient and transcultural aspirations to radically transform human existence, socially and bodily", notwithstanding his insistence that "the transhumanist movement [is] a *modern* form of Enlightenment techno-utopianism [emphasis added]" (p. 757).

figures such as Giovanni Pico della Mirandola (1463–1494 CE), yet fails to identify the connection between the latter and various hermetic and occult currents within European experience. Noble (1997), Davis (1998), Gray (2007), Zimmerman (2008, 2009) and Hughes (2012) trace some of these motivations to technological manifestations of Gnostic, millenarian/millenialist and apocalyptic currents within medieval Western Christianity, and drawing on recent scholarship at the intersection of critical race theory and critical theory of religion (Heng 2011a; Lloyd 2013; Maldonado-Torres 2014a, b), I want to suggest that such "techno-millenialist" currents feed into the emerging "technology" of race at the onset of colonial modernity commencing with the Columbian voyages in 1492 CE, a technology that is being subjected to iterative transformation along transhumanist and subsequently technological posthumanist lines.

4 Transhumanist Demographics

Having established the genealogical links between humanism, transhumanism, technological posthumanism and the European Enlightenment, and having drawn attention to the non-European "backdrop" which facilitated this project, I now turn to consider the demographic constitution of the transhumanist movement in terms of its racial composition.

In this connection, Pellissier (2013) presents some interesting statistics regarding the "ethnic" self-classification of transhumanists, viz. 85.4% white, 3.3% Asian, 1.0% black and 10.0% multiple races.[13] Complementing these findings, a more recent survey of the beliefs held by so-called technoprogressives conducted by Hughes (2017) revealed that only 35% self-identified as anti-racist. In addition to what this figure might indicate vis-à-vis relative lack of engagement with the issue of race among transhumanists, it is not at all clear how racism and anti-racism were understood in Hughes' survey, by questioners and respondents alike: for example, was racism framed in tacitly liberal terms—that is, as something personal, irrational, transient and exceptional—or was it understood along critical race theoretical and decolonial lines as a systemic, rational, persistent and pervasive structural phenomenon (Goldberg 1993)?

Notwithstanding the international nature of transhumanist, extropian and related techno-scientific movements, and granted the need to take seriously the "hybrid" nature of these endeavours involving the contributions of various ethnicities, genders and nationalities, it is empirically demonstrable on demographic grounds, both quantitative and qualitative, that transhumanism is hegemonically white, male and

[13]Notwithstanding concerns about conflating ethnicity with race, some of the responses to the question about ethnicity/race in the survey are quite revealing and include the following: "Human (for now)", "this question is stupid", "race doesn't exist", "Homo Mutantes", "relevance?" and perhaps most telling of all, "aren't we beyond the importance of subspecies distinction?", thereby pointing to a reductively biological conception of race.

"Western" (Euro-American)[14]; furthermore, and as previously argued, it is a project whose trajectory is traceable, genealogically, to a specific historical and geographical experience, viz. the European Enlightenment. On this basis, and in terms of its entanglement with race, I want to suggest that transhumanism should be understood as a Eurocentric/West-centric/white-centric phenomenon.[15] As to whether such hegemony/centrism entails that transhumanism is a racial—more precisely, *racist*—phenomenon, critical theorist Dale Carrico (2012) maintains that "one needs to recall at the outset that one can benefit from racist legacies or mobilize racist discourses without necessarily affirming racist beliefs, indeed while earnestly affirming anti-racist ones, and so recognizing the force of racism is often a matter of exposing structural effects rather than making accusations of unalloyed bigotry". He then goes on to ask whether "'digital-utopian' disdain of 'meat bodies' [is] racist" suggesting that

> It need not be on its face, certainly, but given the whiteness of these subcultures and of the 'reason' qua rage for order with which they so often identify, and given the distressing tendency of race to function precisely as a discourse producing bodies raced qua 'the bodily' as such—that is to say, as the epidermalized body, the muscularized body, the body as bestialized, infantilized, precivilized atavism, the body as seat of irrational and threatening passions, and on and on—one doesn't have to look very hard to find all sorts of racist symptoms cropping up in these precincts.

Yet despite explicit recognition of racism as a "legacy system", notwithstanding his arguably reductive association of race with the body in the above statements, Carrico rather problematically concludes his brief exploration of this question by stating: "Is transhumanism racist? I leave that to the reader to decide." Carrico's ambivalence is somewhat perplexing given his assertion in a later work that

> It grows ever more difficult to shake the troubling analogies between humanism and its debased techno-scientific companion discourse: the race science that legitimized every brutal imperial, colonial, globalizing, ghettoizing, apartheid regime in modern memory. The putative neutrality of the optimal human to which transhumanist enhancement genuflects is obviously another vestige of this parochially raced universal human, post-human though it may be (Carrico 2013, p. 59)

I suggest that it is Carrico's failure to correctly "name" "The World" that leads him to a certain ambivalence vis-à-vis answering the "race question" as it pertains to transhumanism and that his framing of the world system as neo-liberal capitalism

[14]The overwhelming whiteness of transhumanism is readily evidenced through quantitative analysis of the demographics of leading transhumanist organizations such as the Institute for Ethics and Emerging Technologies (IEET) and Singularity University (SU); for example, in 2017, 82% of faculty speakers were identifiable as white (Source: https://su.org/faculty-speakers/).

[15]In this connection, Ferrando (2014) maintains that "within a posthumanist frame, race and its intersections with gender, class, and other categories, have yet to be fully addressed" (p. 13), and that "[there is a] need for a deeper investigation in the topic of race, ethnicity and their intersectional significations in the development of technological futures" (p. 15). What Ferrando fails to consider here is the possibility that transhumanism and technological posthumanism might constitute projects forged *in opposition to* the racialized "other".

rather than religio-racial colonial modernity prevents him from adopting a more definitive stance about the ontological and political consequences of the hegemonic whiteness of transhumanism.[16]

5 Whiteness and "White Crisis"

In framing my argument for transhumanism and/as whiteness, and in pointing to the demographically hegemonic whiteness of the transhumanist movement, there is a need to clarify what is meant by the term *whiteness* beyond its association with epidermal considerations. In this connection, I draw upon the sociological exploration of whiteness due to Garner (2007, 2010a, b)—specifically, (1) his processual understanding of whiteness as a phenomenon existing in dynamic relational-tension to other racialized identities, (2) the function of whiteness as a tacit invisible background standard and (3) the socio-political structural manifestation of whiteness as a persistent, yet contested, globally systemic political structure, viz. white supremacy, a position he derives from Mills (1997). While concerns about the future of whiteness have been engaged by some decolonial commentators against the backdrop of a purported shift to a "post-racial" reality (Alcoff 2015; Sayyid 2010, 2017), anxieties about the future (or otherwise) of whiteness are arguably traceable to the late nineteenth- and early twentieth-century CE phenomenon of "White Crisis"explored by Füredi (1998) and Bonnett (2000, 2005, 2008), the latter of whom refers to a decline of overt discourses of whiteness—more specifically, white supremacism—and the concomitant rise of a discourse about "the West".[17]

[16]Carrico (2013) maintains that "it is the whole terrain of ongoing technodevelopmental social struggle that defines post-humanist strategies and sensibilities, rather than any particular post-human personage, tribe, or social formation thrown up in any one moment of that world-historical technodevelopmental storm-churn. The posthuman need not be a singular imaginary prostheticized personhood eliciting asymptotic approximation via successive enhancements [emphasis added]" (p. 59). Going further he suggests that "it seems … disastrous to conceive post-humanism as a moralizing identification with some tribe defined in its fetishization of idiosyncratic artifacts or techniques, real or imagined. Rather, we should think of post-humanism as an ethical recognition of the limits of humanism provoked by an understanding of the terms of ongoing technodevelopmental social struggle as well as an ethical demand that this struggle always materialize equity-in-diversity" (p. 60). In arguing along these lines, Carrico attempts to differentiate "critical" from "technological" posthumanism; however, I would suggest that it is unclear to what extent a critical orientation can be maintained given the utility of disembodied conceptions of information on account of their wide applicability and facilitative power (Hayles 1999, 2003), the increasing ubiquity of informationalist conceptions of phenomena and the conflation of various forms of critical posthumanism including those conceptualized in informational terms such as the techno-progressive account articulated by Haraway in her *Cyborg Manifesto*. Given the hegemonic whiteness of transhumanism, I would suggest there is a sedimented dispositional bias towards re-articulating "tribal" white supremacy in techno-scientific form at work here.

[17]It is important to note that this "crisis" literature appears at a time when proclamations of "white racial supremacy" are being articulated in public by various commentators belonging to the

It is suggested that the election of Donald Trump as President of the USA, the Brexit phenomenon in the UK and the continued rise of far/alt-right politics in the USA and Europe can—and *should*—be seen as *one* response to the re-emergence of the phenomenon of White Crisis, almost 50 years on from the anti-racist struggles of the 1960s, and almost a century on from when White Crisis was first being discussed in the West (specifically, Britain and America). In this connection, it is crucial to note that according to Bonnett (2008), "whiteness and the West . . . are both projects with an in-built tendency to crisis. From the early years of the last century . . . through the mid-century . . . and into the present day . . . we have been told that the West is doomed" (p. 25).

While there has been a mainstream tendency, on both the right and the left, to frame the above—that is, the election of Trump and the Brexit phenomenon—in economic terms,[18] this reading has been contested by sociologists such as Sayer (2017) and Bhambra (2017), the first of whom offers a penetrating analysis of the demographic statistics associated with US and UK polling and voting patterns. Against economistic readings, Sayer points to the role of nostalgia, the loss of identity and the rise of authoritarian populism driven by racist, xenophobic and nativist sentiment in the wake of immigration, all of which need to be considered against the rising backdrop of more overt and crude manifestations of "far-right" nationalism. Crucially, in this connection, Sayer goes as far as to assert that "*this is*

---

dominant Euro-American powers of the1920s and 1930s. Less than 50 years on, whiteness (under the signifier "the West") is once again ostensibly facing crisis as a result of the Civil Rights Movement in the USA and the increasing linkage of this struggle to global anticolonial struggles. Formal independence from European colonial powers is achieved in the late 1960s, and the Civil Rights struggle achieves certain limited victories; however, structures of colonial domination persist in the "operating logics" of the newly independent postcolonial states, decolonization as a project arguably being aborted under the transition from a liberal to a neo-liberal world order in the 1980s onwards. As neo-liberalism morphs into neo-conservativism, the "apocalyptic" project of a "war on terror" surfaces (Gray 2007) and the historically sedimented figure of the Muslim "other" as threat/enemy *re*-emerges (Ali 2017a). Yet, concurrent with and at least partly due to this centring of the specifically *Muslim* "other" as enemy and the need to mobilize for war against it, "breathing room" is provided in South America, South Asia and latterly South Africa for the gestation and development of a *decolonial* project—that is, *re*-engagement with the unfinished project of decolonization (to be contrasted with Habermas' unfinished project of modernity). During the 2000s, the "decolonial option" begins to be embraced by some members of "minority" non-white groups located in the West, this tendency escalating in the post-racial era under Obama, with increasing contestation of whiteness and Eurocentrism in the academy, activist mobilizations against anti-blackness and white supremacy in movements such as Black Lives Matter and various contemporary anti-racist responses to the rise of the far/alt-right in the USA and Europe against the backdrop of the continued rise of Islamophobia.

[18]Offering a more nuanced economic reading focusing on the fallout from financial crises vis-à-vis entanglement with "race matters", Gupta and Virdee (2018) maintain that "the [2007–2008] financial crisis incorporated and generated a web of crises gripping social life at various levels" (p. 1749). In arguing that such webs should be seen as *generated* by the financial crisis, yet also *incorporating* such webs, I want to suggest that the financial crisis be situated against a broader backdrop of White Crisis which arguably has long durée entanglements with European religio-racial apocalypticism; on this point, see (Ali 2019).

centrally a race war—a war on the ethnic Other, be it a Black (lives matter), Syrian (refugee), Mexican, Polish, Chinese, or Muslim Other—that has successfully managed to pass itself off as a revolt of the deprived and the dispossessed" (p. 102). Consistent with Sayer's line of critique, Bhambra presents an equally important statistically informed critical analysis of how the white working class has been mobilized with a view to providing an economic explanation for the Trump and Brexit phenomena, a rhetorical move which functions to obscure the reality that in both cases the *decisive* cause lies with the white middle class. According to Bhambra, "the skewing of white majority political action as the action of a more narrowly defined white working class served to legitimize analyses that might otherwise have been regarded as racist" (p. 214). On her view, "a focus on who actually voted for Brexit (and Trump) would reveal that the opposition to immigration was primarily cultural in character and not based on economic disadvantage. It extended beyond a white working class to include the white middle class (p. 222). Crucially, in relation to the argument present herein, Bhambra maintains that both phenomena need to be understood in terms of

> Gradually *decreasing inequality* rather than a decline in the conditions of the white middle class . . . the white middle class, believes that these newly experienced material conditions of *greater equivalence* are not appropriate to their place [emphases added] (p. 226)

In short, the issue is relational rather than substantial insofar as it turns on non-white contestation of "white privilege" and is a manifestation of White Crisis.

6 Algorithmic Racism

In terms of thinking specifically about *transhumanism* and/as whiteness, I want to argue that transhumanism/technological posthumanism should be viewed as a somewhat *different* response to the phenomenon of White Crisis, one that is techno-scientific and occurs in parallel with, albeit being somewhat obscured by, the more overt phenomenon of conservative/middle-class "White Backlash" vis-à-vis the socio-political phenomena described above.[19] In particular, I want to argue that the shift described by Füredi and Bonnett from white to West is usefully framed in terms of the re-inscription[20]—or rather, "algorithmic" *re-iteration*—of whiteness under different signifiers including the techno-scientific signifier of transhumanism

[19]Such phenomena include the resurgence of strident and protectionist "strong-man" nation-statism, racialized articulation and foregrounding of "concerns" about border controls, immigration, citizenship and notions of "belonging", along with the rise of cruder and more overt forms of white supremacy in comparison with what was arguably the more subtle, more refined and more covert operation of the socio-political logics of "racial liberalism" (Goldberg 1993; Mills 1997) in Western nation state formations.

[20]Bonnett (2008) appears to concede the "iterativity" of whiteness in referring to its "re-invention", "well into the twenty-first century", pointing out that "the history of whiteness is one of transitions and changes" (p. 17).

associated with the convergence of GRIN/NBICS technologies; furthermore, that this shift in whiteness needs to be situated within a longer historical frame than that going back to the late nineteenth century CE, arguably one that commences with the Columbian voyages in 1492 CE which resulted in the emergence of a racialized world system (Ali 2017a)[21]; moreover, a history involving *other* paradigmatic[22] shifts including those from religious to philosophical to scientific and latterly cultural expressions of race/racism/racialization, such transformations constituting re-articulations of the difference between the human (European) and the sub-human (non-European).[23] Insofar as such iterations might be seen as different manifestations of the same phenomenon—thereby pointing to a certain continuity through change—I want to suggest that they are usefully understood in terms of what has been described elsewhere as "algorithmic racism" (Ali 2016b, 2017b).[24] Algorithmic racism (AR) postulates the existence of a historically contingent, yet sedimented and dispositional, "meta-process" linking racialization processes, and is motivated by a concern to assist with the disclosure of continuities that are masked (obscured, occluded) by transitions between different materializations—that is, iterations—of race/racism in different historical epochs, and *a fortiori* in the transition from colonial modernity to the contemporary postmodern/postcolonial era. I argue that

[21]In this connection, it is interesting to note that according to Hayles (2003), "we do not leave our history behind but rather, like snails, carry it around with us in the sedimented and enculturated instantiations of our pasts we call our bodies" (p. 137). Apart from the need to decolonially interrogate *whose* history needs to be considered (body-politics of knowledge) and from where (geo-politics of knowledge) in her invocation of the first person plural ("we"), it is crucial to note that it is not *just* bodies that are sites for sedimentation and enculturation, but regimes of governmentality which include but transcend the body so as to incorporate institutions, land, discursive practices, etc.; on this point, see Hesse (2007).

[22]The qualifier "paradigmatic" is necessary in order to draw attention to the entangled relationship between prior and posterior dominant (or "signature") articulations of race evincing the interplay of the sequential, the decisional and the iterative within algorithmic racism wherein different iterations of race/racism/racialization are marked by paradigmatic shifts in formation that should be considered distinct, but not oppositional, and at least partly inclusive rather than wholly exclusive. For a useful discussion of such entangled racialized logics, see Heng (2011b, c).

[23]What tends to be obscured, if engaged at all, in discussions of the relationship between the human and the transhuman is the prior relationship between the human and the sub-human (which should not be conflated with the broader category of the *non*-human), the latter providing the ontological ground against which the former is constituted through a process of hierarchical negative dialectical opposition, viz. the human (superior) as the negation of the sub-human (inferior).

[24]In this connection, it should be noted that there are precedents for thinking about race as algorithmic. For example, Coleman (2009) maintains that "race as we know it is an 'algorithm' inherited from the age of Enlightenment" (p. 184). While concurring that race *can* be understood as an algorithm, I want to suggest that dating this phenomenon—or at least its heritability—to the Enlightenment is problematic, and that race emerges much earlier, in systemic/structural form during the long durée of the sixteenth century CE (and for critical medievalists, to the Middle Ages). In addition, I want to suggest that not only is race an algorithm, metaphorically speaking, but that following the "cybernetic turn" of the 1950s, the continued rise of informational, computational and algorithmic logics (technical, social, cultural, economic, political, etc.) has resulted in a situation wherein the racial algorithm has engendered algorithmic formations of race.

the contemporary moment is marked by a relational shift *from* the distinction between sub-human (non-European, non-white) and human (European, white) *to* that between human (non-European, non-white) and transhuman (European, white), such iterative transformation being prompted, at least partly, by various non-European/non-white contestations of Eurocentric/West-centric/white-centric conceptions of the human against the much broader background or horizon of a resurfacing of the phenomenon of White Crisis.[25]

7 Conclusion

In this essay, I have argued that transhumanism *and* whiteness should be considered in relation to each other—more precisely, that transhumanism should be understood *as* whiteness when conceptualized in terms of the operation—or "execution"—of the iterative logics of algorithmic racism. I have further suggested that this line of reasoning should be understood against the backdrop of a resurfacing of the phenomenon of White Crisis, and that transhumanism and technological posthumanism are usefully understood as techno-scientific responses to this phenomenon.

References

Alcoff, L.M.: The Future of Whiteness. Polity Press, Cambridge (2015)

Ali, S.M.: Towards a critical race theory of information. In: The Fourth ICTs and Society Conference: Critique, Democracy, and Philosophy in 21st Century Information Society – Towards Critical Theories of Social Media. Uppsala University, Sweden, May 2–4, 2012. Abstracts booklet, pp. 58–59. www.icts-and-society.net/wp-content/uploads/Abstracts.pdf (2012)

Ali, S.M.: Race: the difference that makes a difference. tripleC. **11**(1), 93–106 (2013)

Ali, S.M. (2014) Towards a decolonial computing. Elizabeth A. Buchanan, de Laat Paul B, Herman T. Tavani and Jenny Klucarich Ambiguous Technologies: Philosophical issues, Practical Solutions, Human Nature: Proceedings of the Tenth International Conference on Computer Ethics – Philosophical Enquiry (CEPE 2013). Porto, Portugal: International Society for Ethics and Information Technology, pp. 28–35

Ali, S.M.: Orientalism and/as information: the indifference that makes a difference. DTMD 2015: 3rd International Conference. In: ISIS Summit Vienna 2015 – The Information Society at the Crossroads, 3–7 June 2015, Vienna, Austria. https://doi.org/10.3390/isis-summit-vienna-2015-S1005 (2015)

Ali, S.M.: A brief introduction to decolonial computing. XRDS, Crossroads, the ACM Magazine for Students – Cultures of Computing. **22**(4), 16–21 (2016a)

---

[25] In this connection, consider the significance of Zimmerman's (2008) highly perceptive observation that "posthumanists often regard humans as *relay runners* about to pass the baton to oncoming others, who in turn will race toward a summit that surpasses all ordinary human understanding [emphasis added]" (p. 363).

Ali, S.M.: Algorithmic racism: a decolonial critique. In: 10th International Society for the Study of religion, Nature and Culture Conference, 14–17 January 2016, Gainesville, FL (2016b)
Ali, S.M.: Islam between inclusion and exclusion: a (decolonial) frame problem. In: Hofkirchner, W., Burgin, M. (eds.) The Future Information Society: Social and Technological Problems, pp. 287–305. World Scientific, Singapore (2017a)
Ali, S.M.: Decolonizing Information Narratives: Entangled Apocalyptics, Algorithmic Racism and the Myths of History. DTMD 2017: 6th International Conference. In: IS4IS Summit Gothenburg 2017 – Digitalisation for a Sustainable Society, 12–16 June, Gothenburg, Sweden. http://www.mdpi.com/2504-3900/1/3/50 (2017b)
Ali, S.M.: White crisis' and/as 'existential risk': the entangled apocalypticism of artificial intelligence. Zygon J. Rel. Sci. **54**(1), 207–224 (2019)
Badmington, N.: Theorizing posthumanism. Cult. Crit. **53**, 10–27 (2003)
Bhambra, G.K.: Brexit, trump, and 'methodological whiteness': on the misrecognition of race and class. Br. J. Sociol. **68**(S1), S214–S232 (2017)
Bonnett, A.: Whiteness in crisis. Hist. Today. **50**(12), 38–40 (2000)
Bonnett, A.: From the crises of whiteness to western supremacism. ACRAWSA (Australian Critical Race and Whiteness Studies Association). **1**, 8–20 (2005)
Bonnett, A.: Whiteness and the west. In: Dwyer, C., Bressey, C. (eds.) New Geographies of Race and Racism, pp. 17–28. Ashgate, Aldershot (2008)
Bostrom, N.: A history of transhumanist thought. In: Rectenwald, M., Carl, L. (eds.) Academic Writing Across the Disciplines. Pearson Longman, New York (2011)
Bostrom, N.: Introduction – the transhumanist FAQ: a general introduction. In: Mercer, C.R. (ed.) Transhumanism and the Body: The World Religions Speak, pp. 1–17. Palgrave Macmillan, New York (2014)
Carrico, D.: 'Is transhumanism racist?' AMOR MUNDI blog. 21 December 2012. https://amormundi.blogspot.co.uk/2012/12/is-transhumanism-racist.html (2012)
Carrico, D.: Futurological discourses and Posthuman terrains. Existenz. **8**(2), 47–63 (2013)
Chun, W.H.K.: Race and/as technology; or, how to do things to race. Camera Obscura. **24**(1), 6–35 (2009)
Coleman, B.: Race as technology. Camera Obscura. **24**(1), 177–207 (2009)
Davis, E.: Techgnosis: Myth, Magic and Mysticism in the Age of Information. Harmony Books, New York (1998)
Ferrando, F.: Posthumanism, transhumanism, antihumanism, metahumanism, and new materialisms: differences and relations. Existenz. **8**(2), 26–32 (2013)
Ferrando, F.: Is the post-human a post-woman? Cyborgs, robots, artificial intelligence and the futures of gender: a case study. Eur. J. Futures Res. **2**, 43 (2014). https://doi.org/10.1007/s40309-014-0043-8
Füredi, F.: The Silent War: Imperialism and the Changing Perception of Race. Pluto Press, London (1998)
Garner, S.: Whiteness: An Introduction. Routledge, London (2007)
Garner, S.: Racisms: An Introduction. Sage, London (2010a)
Garner, S.: White Identities: A Critical Sociological Approach. Pluto Press, London (2010b)
Goldberg, D.T.: Racist Culture. Blackwell, Oxford (1993)
Gray, J.: Black Mass: Apocalyptic Religion and the Death of Utopia. Penguin, London (2007)
Gupta, S., Virdee, S.: Introduction: European crises: contemporary nationalisms and the language of 'race. Ethn. Racial Stud. **41**(10), 1747–1764 (2018)
Hayles, N.K.: How We Became Posthuman: Virtual Bodies in Cybernetics, Literature, and Informatics. Chicago University Press, Chicago, IL (1999)
Hayles, N.K.: The human in the posthuman. Cult. Crit. **53**, 134–137 (2003)
Heng, G.: Holy war Redux: the crusades, futures of the past, and strategic logic in the 'clash' of religions. PMLA. **126**(2), 422–431 (2011a)
Heng, G.: The invention of race in the European middle ages I: race studies, modernity, and the middle ages. Lit. Compass. **8**(5), 315–331 (2011b)

Heng, G.: The invention of race in the European middle ages II: locations of medieval race. Lit. Compass. **8**(5), 332–350 (2011c)
Hesse, B.: Racialized modernity: an analytics of white mythologies. Ethn. Racial Stud. **30**(4), 643–663 (2007)
Hughes, J.J.: The politics of transhumanism and the techno-millennial imagination, 1626–2030. Zygon. **47**(4), 757–776 (2012)
Hughes, J.J.: What do technoprogressives believe? IEET Survey 2017. http://ieet.org/index.php/IEET/more/hughes20170218 (2017)
Islam, M.: Posthumanism and the subaltern: through the postcolonial lens. http://indiafuturesociety.org/posthumanism-subaltern-postcolonial-lens/ (2014)
Jotterand, F.: At the roots of transhumanism: from the enlightenment to a post-human future. J. Med. Philos. **35**, 617–621 (2010)
Lloyd, V.: Race and religion: contribution to symposium on critical approaches to the study of religion. Crit. Res. Rel. **1**(1), 80–86 (2013)
Mahootian, F.: Ideals of human perfection in sufism and transhumanism: a comparison. In: Tirosh-Samuelson, H., Mossman, K.L. (eds.) Building a Better Human? Refocusing the Debate on Transhumanism, pp. 133–156. Peter Lang Press, Berlin (2012)
Maldonado-Torres, N.: AAR centennial roundtable: religion, conquest, and race in the foundations of the modern/colonial world. J. Am. Acad. Relig. **82**(3), 636–665 (2014a)
Maldonado-Torres, N.: Race, religion, and ethics in the modern/colonial world. J. Rel. Ethics. **42**(4), 691–711 (2014b)
Mignolo, W.D.: The Darker Side of Western Modernity: Global Futures, Decolonial Options. Duke University Press, Durham (2011)
Mills, C.W.: The Racial Contract. Cornell University Press, Ithaca (1997)
Moore, M.: The philosophy of transhumanism. In: More, M., Vita-More, N. (eds.) The Transhumanist Reader: Classical and Contemporary Essays on the Science, Technology, and Philosophy of the Human Future, 1st edn, pp. 3–17. Wiley, New York (2013)
Noble, D.F.: The Religion of Technology: The Divinity of Man and the Spirit of Invention. Penguin, New York (1997)
Ogunnaike, O.: From heathen to sub-human: a genealogy of the influence of the decline of religion on the rise of modern racism. Open Theol. **2**, 785–803 (2016)
Pellissier, H.: Transhumanists: who are they, what do they want, believe and predict? J. Pers. Cyberconsciousness. **8**(1), 20–29 (2013)
Sayer, D.: White riot—Brexit, trump, and post-factual politics. J. Hist. Sociol. **30**, 92–106 (2017)
Sayyid, S.: Do Post-Racials Dream of White Sheep? Tolerance Project Working Paper, Centre for Ethnicity and Racism Study, pp. 1–14. Leeds University, Leeds (2010)
Sayyid, S.: Post-racial paradoxes: rethinking European racism and anti-racism. Patterns Prejudice. **51**(1), 9–25 (2017)
Sombetzki, J.: How 'post' do we want to be – really? The boon and bane of enlightenment humanism. Cultura Int. J. Cult. Axiol. **13**(1), 161–180 (2016)
Zimmerman, M.E.: The singularity: a crucial phase in divine self-actualization? Cosmos Hist. J. Nat. Soc. Philos. **4**(1–2), 347–370 (2008)
Zimmerman, M.E.: Religious motifs in technological posthumanism. West. Humanit. Rev. **3**, 67–83 (2009)

Chapter 13
Promethean Shame Revisited: A Praxio-Onto-Epistemological Analysis of Cyber Futures

Wolfgang Hofkirchner

Abstract In this chapter, imaginaries of the future with respect to cyber technologies will be analysed. The question is whether or not the relationship between humans and machines shall be designed, modelled and framed such that the distinction between the human and the artificial is blurred. Answers that enact conflating (reductive or projective) or disjoining ways of thinking give evidence of different combinations of hubris and humiliation. For philosopher Günther Anders, in the 1950s, "Promethean shame", that is, hubristic self-humiliation, was the "climax of all possible dehumanization". Today, this anti-humanism comes in trans- and posthumanist disguises. Only integrative answers without hubris and without humiliation can provide humanist imaginaries.

1 Introduction

Whilein the USA around the time of WWII, philosopher Günther Anders witnessed and was deeply affected by the achievements of advanced industrial production and emerging "big science" (Bernal 1986). This scientific-technological revolution ushered in the atomic age and awareness of the capacity for mass destruction. The period also kicked off the development and diffusion of computers, and "informatization" (Nora and Minc 1978), leading to the information age. In his essay "Über prometheische Scham" published in his *Die Antiquiertheit des Menschen*, Anders first formulates an argument that reads as a forerunner of contemporary critique of anti-humanist positions (Anders 1956). These positions go hand in hand with technological trends that seem to manifest a blurring of the distinction between the human and the artificial, and which are also woven into current trans- and

W. Hofkirchner (✉)
The Institute for a Global Sustainable Information Society (GSIS), Wien, Austria

Institute of Visual Computing and Human-Centered Technology, TU Wien, Austria
e-mail: wolfgang.hofkirchner@gsis.at

W. Hofkirchner, H.-J. Kreowski (eds.), *Transhumanism: The Proper Guide to a Posthuman Condition or a Dangerous Idea?*, Cognitive Technologies,
https://doi.org/10.1007/978-3-030-56546-6_13

posthumanist imaginaries. As such, it seems worth reprising Anders's argument in order to shed light on current positions.

Anders claims that humans experience "Promethean shame" in response to things fabricated by human artifice. Anders first presented his argument in 1942 at a Los Angeles seminar convened by the Frankfurt School critical theorists Theodor Adorno and Max Horkheimer. The seminar was also attended by Bertolt Brecht and Herbert Marcuse. Anders subsequently claims:

- "Artificiality is the nature of human beings. This means that the demand human beings place on the world from the outset surpasses what it supplies" (quoted in Müller 2016, 99).
- "[...] humans become the product of their own products" (Müller, 100).
- "[...] human beings can no longer live up to the demands that their own products place upon them" such that "[a] discrepancy, a widening gulf opens between the human and its products [...]" (ibid).

These steps are the basis for Promethean shame. After Copernicus, Darwin and Freud, Anders describes here another blow dealt to humanity's sense of itself. He provides a striking illustration in section 10 of his essay. US General Douglas MacArthur, "at the beginning of the Korean War proposed measures that arguably could have triggered a Third World War. [...] the decision as to whether such an outcome should be risked or not was taken out of his hands. Those who removed his responsibility from him, however, [...] removed the decision [...] to hand it over to a *machine*", "to an '*electric brain*'" (Anders 2016, 58).

The "electronic brain" opted against MacArthur's approach, but Anders's emphasis is on the fact that "the process, as such, by which this decision was reached, was at the same time the most epoch-making defeat that humanity could have inflicted on itself. For never before had humanity degraded itself to such a degree that it entrusted judgement about the course of history, perhaps even about whether it may be or not be, to a thing" (60–61). Anders qualifies Promethean shame as "*hubristic self-humiliation*" (Anders 1956, 47, translation by W.H.). Anders argues in critique that according to such an attitude man is deemed a faulty construction but at the same time capable of constructing artefacts that promise to transcend the faults of the human condition, and so offer completions or perfections, but of a particular quality; for Anders, this development is the "*climax of all possible dehumanization*" (Anders 2016, 44).

In this chapter, I initially explore how hubris and humility are in varying ways implied by a diversity of human–machine relationships. That is, in practice, in knowledge and in the process of producing knowledge. My argument supports the critique of anti-humanist future imaginaries and contests the blurred distinction between the human and the artificial while also questioning the realism of inferred trends and the methods used to justify positions.

My argument makes use of praxio-onto-epistemology. This approach contends that praxiology, ontology and epistemology build levels in philosophy, with praxiology as the uppermost level. Praxiology provides answers to the questions: what is the reason for which we want to intervene in the world, what are the means by which we intervene and what are the results of our interventions? Ontology provides answers to the question of how the world works (helping us understand how we can successfully intervene in the world), and epistemology addresses the question of which tools we might use to acquire knowledge (in order to successfully understand how the world works). Praxiology, ontology and epistemology form a hierarchy:

> There is relative autonomy of each of the domains (praxis may shape reality but reality provides the scope of possible practices; reality may shape method but method provides the scope of possible realities) and of each of the disciplines (praxiology does not fully determine ontology and ontology does not fully determine epistemology, and *vice versa*) (Hofkirchner 2013, 47–48).

The relevant context for our purposes in this chapter is imaginaries of future societies shaped by (computerized) information (and communication) technologies and artefacts:

- The praxiological aspect concerns the relationship of the human and the artificial, "man" and "machine", during the process of designing social systems and technologies.
- The ontological aspect manifests in the (theoretical) modelling of the relationship in 1.
- The epistemological aspect comes to the fore when methods frame the investigation of man and/or machine and contribute to the modelling of their relationship in 2.

Following a praxio-onto-epistemological approach, it is also useful to consider the different ways of thinking that characterize underlying assumptions concerning how practices, entities/events or phenomena are interrelated. There are, in principle, only a few possibilities concerning interrelations: identity of units without regard to difference, difference without regard to identity and integration of identity and difference.

Identity of units may be the result of conflation. Conflation happens in two ways: reductionism or the levelling down of a higher to a lower unit pertaining to their complexity, and projectionism or the levelling up of a lower to the higher unit. Difference may express disjunction: a higher unit may be stated to be distinctive over and against a lower one, a lower unit may be stated to be distinctive or both may be stated to be so, leading, finally, to overall indifference. Integration implies that units are connected according to their degree of complexity. The lower is included in, and implied by, the higher and the lower shares with the unit that has a higher degree of complexity at least one property, but the higher unit is in the exclusive possession of at least another property (so there are degrees of identity and difference).

In the context of imaginary cyber futures:

- Reductionist ways of thinking appear as *technomorphic.*[1] That is, identifying man and machine, according to their technical forms.
- Projectionist ways of thinking appear as *anthropomorphic.*[2] That is, identifying man and machine according to their social forms.
- Disjunctive ways of thinking appear as *anthropocentric,*[3] *technocentric* or *interactivistic*. That is, differentiating man and machine according to their respective social, technical or independent forms;
- Integrative ways of thinking appear as *technosocially systemic*. That is, identifying and differentiating man and machine as parts of a common whole.

In the rest of the chapter, I set out an ordered schematic structure of these in relation to praxiology (the design of technology), ontology (theoretical models) and epistemology (methodological framing) using Anders's work as a point of departure. I do so in three extended sections: section 1, hubris and/or humiliation; section 2, agents and/or patients; and section 3, mono- and/or multi-/interdisciplinarity. In each section I argue, because of shortcomings of three of the four ways of thinking, towards a final subsection setting out a preferred systems science position that is praxiologically neither hubristic nor humiliating but synergistic, ontologically neither agential nor patient-centred but emergentist and epistemologically neither mono- nor multi- or interdisciplinary but transdisciplinary.

2 Praxiology: Hubris and/or Humiliation?

Man–machine praxiology concerns the practices collective actors deploy when designing social systems by technological means. Such practices reveal a logic that, willingly or not, expresses a certain way of thinking regarding how man and machine relate. There are three relevant logical types: a universal colonization, a particularistic compartmentation (as the plain negation of colonization) and synergistic nesting (as a negation of both universalism and particularism).

2.1 Colonized Man–Machine Designs: Technomorphic and Anthropomorphic Conflation of Practices

According to universalistic practice, the human and the artificial are (to be) treated indiscriminately. Universalistic practice conflates the distinction between treating

[1]The term "technomorphic" is chosen in analogy to the term "anthropomorphic".

[2]The term "anthropomorphic" embraces the human as inherently social. So, "sociomorphic" would be a term that is as good as "anthropomorphic".

[3]The term "anthropocentric" includes the social too.

Table 13.1 Colonized man–machine designs

	"Man–machine" designs			
Conflation	**Colonization:** The human and the artificial shall be treated indiscriminately	**Reduction**	**Technomorphism**	
			Man treated like a machine	**"Homo deus":** Hubris from humiliation (transhumanism)
		Projection	**Anthropomorphism**	
			The machine treated like humans	**"Techno sapiens" (humanoids):** Humiliation from hubris

the human and treating the artificial in one of two ways. Either human activities are colonized by artificialities or vice versa (see Table 13.1).

2.1.1 Technomorphic Colonization: Aspiring to "Homo Deus"

Technomorphic universalism treats the human, or contends that the human shall be treated, like a machine. The praxiological meaning of technomorphism is that the form of the treatment of the human is derived from the form of the treatment of the artificial. This is a reduction in sofaras the human, including her social (cultural, political, economic) world, is more complex than that of the artificial.

Technomorphic universalism encompasses the concept of the cyborg, but in aspirational terms is expressed through "homo deus", a longing to perfect the species through artificial means (homo deus is associated with the work of historian Yuval Noah Harari 2016), which is the essence of transhumanism. As Peter Fleissner also notes, reference to the Internet often offers the potential to realize omniscience, omnipresence, omnipotence and omnibenevolence (Fleissner and Hofkirchner 1998).

This realization of god-like qualities or perfection is to be obtained through mechanistic interventions and alterations of the human body, manipulated as an object. Identified "shortcomings" of the human condition are social problems to be solved technically in contrast to the cultivation of social abilities and ambitions of humankind. So, hubris is expressed through the desire to become god-like, but this is conjoined with Anders's humiliation, in the sense that what is human is degraded and reduced to the utilization of a narrow spectrum of technological enhancement and augmentation opportunities. Such a technomorphic treatment, which derives hubris from humiliation, seems undesirable.

2.1.2 Anthropomorphic Colonization: Aspiring to "Techno Sapiens"

Anthropomorphic universalism reverses the technomorphic man–machine relation. Such universalism treats the machine, or contends that the machine shall be treated, like the human. Anthropomorphism means that the form of the treatment of the artificial derivatives from the form of the treatment of the human. Following the reverse logic of the previous section, treating machines as (yet-to-be-developed) humans is a projection.

Anthropomorphism consists in attempts to perfect the artificial by making it more and more human-like, that is, humanoid robots, androids, etc., culminating in what Peppo Wagner refers to as "techno sapiens" (2016): a machine that is purportedly equivalent to the human, at least in terms of parameters taken to be essential to establish that the entity is indistinguishable from the human. This trend is not new. It has, since the beginning of informatization, been recognizable within the language used to describe features of computing, artificial intelligence and robots (consider the derivation of "computer", "robot" and the connotation of "man–machine communication" and so forth). It is also inscribed in the mass production processes that introduced fixed automation and then flexible automation, and in so-called autonomous systems. It is identifiable in debates concerning whether or not moral rules should be implemented in and for robots; though it might be argued, as philosopher Susanne Beck points out, that machines cannot rationalize on the basis of sentience, values and intuition, and cannot be taught the evaluation of infinitely contingent situations, and thus cannot act morally even if they could mimic human decisions (Stanzl 2017). This is despite attempts to devise such an algorithm, for example, by neuro-informatics researchers at the University of Osnabrück (see also Trappl 2015).[4]

Various related issues have arisen, for example, whether or not "acting" and "thinking" machines should be endowed with electronic "personhood" and granted rights and obligations. This possibility is raised by the Commission on Civil Law Rules on Robotics (2015/2103(INL)) and is proposed by a report published by the European Parliament's Committee on Legal Affairs (2017). A resolution was passed by the European Parliament with nearly two-thirds of the votes in favour on 16 February 2017. The issue has an economic and social context. For example, whether or not children and the elderly should be treated by robots as (cheaper and available) substitutes for interaction and communication with caring human persons. Behind this issue lies the further consideration of whether androids could and should be endowed with the capacity to detect emotions in humans, and then not only simulate those emotions (as-if-emotions) but actually have them.

It is worth noting that argument regarding human intelligence tends to be dominated by positivism and behaviourism and this influences what it means for entities to be equivalent. For example, in the Turing test, two entities are declared identical if they behave in the same way, and so cannot be distinguished by an interlocutor. However, this behavioural test does not indicate anything directly about the nature of the entities in question, since differences may be at the ontological level.

[4]See https://ikw.uni-osnabrueck.de/en/ni

Table 13.2 Compartmentalized man–machine designs

	"Man–machine" designs		
Disjunction	**Compartmentation:** The human and the artificial shall be treated in discriminate ways	**Anthropocentrism**	
		Man treated better than machines	**Pride of creation:** Humble hubris
		Technocentrism	
		The machine treated better than men	**Übermensch ex machina:** Hubristic humility (posthumanism)
		Man–machine interactivism	
		Both treated on equal terms	**Hybrid networks:** Hubris-humility shifting

Returning to Anders, hubris can be identified here because of the intention to create artificial humans and the conviction that this is possible. This represents a humiliation insofar as these artificial humans dictate to (in given circumstances as above) "natural" humans. Thus, while technomorphism derives hubris from humiliation, anthropomorphism derives humiliation from hubris. This seems no more desirable than the previous case.

2.2 *Compartmentalized Man–Machine Designs: Anthropocentric, Technocentric and Interactivistic Distinctions of Practices*

The negation of universalism holds that the human and the artificial are or shall be treated as distinct. This way of thinking yields particularism in practice, since it separates and disjoins the human and artificial. As stated in the introduction, the practice or position that social and technological forms cannot be intrinsically tied to each other and are extrinsic takes three guises: anthropocentric, technocentric and interactivistic (see Table 13.2).

2.2.1 Anthropocentric Compartmentation: Appreciating Human Status as "Pride of Creation" Versus Depreciating "Trumpery"

The first disjunction treats the human as distinctive, such that the human is or shall be perfected independently from the artificial, and where humans and/or society are or shall be designed without resorting to technology. Since this position excludes practice concerning artefacts from its social perspective, it might be called anthropocentric compartmentation. Its core is the conservative belief in human society as "pride of creation" and is rooted in some ideological theocratic beliefs, such as those of the Amish. Distinctiveness, as human exceptionalism, places social processes over and against technological processes and the latter are treated pessimistically as mere trumpery. From this point of view, man/society is highly valued and engineering (science) lacks value or may even be dangerous.

Anthropocentric compartmentation clearly qualifies as hubris, albeit a humble hubris, since it foregoes opportunities technological processes can provide. Again, this position seems undesirable.

2.2.2 Technocentric Compartmentation: Creating the "Übermensch" Ex Machina Versus Deploring "Human Obsolescence"

The second disjunction privileges the artificial, such that the machine is or shall be treated without consideration of man, and where the artificial is perfected independently. The contradicting evaluations of man/society and technology are reversed and dealing with the human in its own right is excluded and so this kind of particularism and compartmentation can be categorized as technocentric. Technological progress is put first.

Technology itself will unavoidably turn into a superhuman, via what is commonly termed the singularity. The mantra of singularitarianism is: such an "Übermensch" will render *Homo sapiens* obsolete. Its point of departure is the assumption that technology runs away, while society's attempts to catch up are doomed to failure. What is feasible will or shall be realized. The development of technology is always ahead of the development of society. Technology is conceived as (a separate) extension of human capabilities such that technological constructs are able to outperform humans. The ideology of posthumanism takes for granted that there is no limit to the perfection of technologies and this leaving behind of humans.

However, technocentric particularism is, returning to Anders, a double-edged sword. On the one hand, it is hubris to believe humans capable of constructing machines that are superhuman. On the other hand, such a superhuman construct would be a self-humiliation of humans. Thus, Anders talks of "*arrogant self-degradation*" and "*hubristic humility*" (Anders 2016, 49). "Promethean shame" follows. In order to be able to become a master, man must turn himself into a slave. Humans refuse "to honour themselves", because the "presumptuously self-aggrandising ideas of entitlement [...] are so aggrandising that they begin to feel inadequate themselves", because "they chastise themselves on account of their own 'backwardness' and the 'shame of having been born'" (Anders 2016, 50).

The resulting position on technological development and the ideology behind it represent the purest case of anti-humanism, and so the least desirable position so far with reference to hubris and self-humiliation.

2.2.3 Interactivistic Compartmentation: Assembling Humans and Machines into Hybrid Networks

The third disjunction does not prioritize either side; human and machine are or shall be treated on equal terms. It is practised or accepted that both humans and artefacts be treated according to their respective forms, in their own right. However, this is founded on indifference rather than a relationship in which both treatments are reconciled at a higher level. The consequence is that the differentiated human and artefact are conflated in terms of the interplay of social and technological practices. To fit the condition of equal terms, differences in dealing with the social and the technological are levelled out. This is epitomized by Bruno Latour's actor-network theory—ANT (2006. Interactivistic compartmentation is also exhibited by the new materialism, an agential realism, of Karen Barad (2007) and Lucy Suchman (2007), where agency is distributed among sociomaterial practices.

Interactive particularism meanders between hubris and humiliation. Hubris comes to the fore when the role of artefacts is equalled to (projected onto) the role of humans, while humiliation comes to light when the role of humans is equalled (reduced) to the role of artefacts. In short, we have got a hubris-humility-shimmering—not a desire at all.

2.3 *Synergistic Man–Machine Design: Technosocial System Integration of Practices for the Good Society*

The negation of reductionist and projectionist universalist ways of thinking and of particularist disjunction leads to integration and synergism. Synergy is an emergent product of a self-organizing system, an added value that is provided to its elements, which binds them together, allowing systems to sustain (Corning 1983). Synergy means a certain constellation of organizational relations that constrains and enables the interaction of the elements, becoming co-action in which every element finds its proper place (Hofkirchner 2017a, 9). Accordingly, in a synergistic man–machine design, the human and the artificial shall be treated "appropriately" (see Table 13.3).

I have previously argued that any social system is a social system by virtue of organizational relations of production and provision of the common good

Table 13.3 Synergistic man–machine design

	"Man–machine" designs		
Integration	**Nesting:** the human and the artificial shall be treated according to different nesting levels	**Technosocial systemism**	
		Man and machine are treated appropriately: technologies are functionalized for the synergy of human actors and social systems	The **"good society":** no hubris no humiliation (*alter-humanism harnessing tools for conviviality*)

(Hofkirchner 2017b). On this basis, the commons are the manifestation of synergy effects. Organizational relations of production and provision determine how the human and the artificial are related. Technology, if meaningful, is oriented towards the advancement of the commons. The advancement of the commons promotes the flourishing of human actors. Thus, artificial devices can and should be nested in the social system, advancing the commons. Informatization needs to be harnessed: There must be reflection—through integrated technology assessment and technology design—of its social usefulness. That is, its expected usage needs to be assessed vs. its actual usage and the consequences thereof; this includes a reflection of both the aptness for the purpose (the utility) and the purpose itself (the function the technology serves).

Appropriate synergism in design overcomes hubris as well as humiliation. In good society, the human and the artificial are related according to the different roles they are given to fulfil in a sustainable system. This provides a long-term corrective in order to guide social and technological development. Arguably, and with reference to Anders, the absence of hubris and humiliation is the only desirable position.

3 Ontology: Agents and/or Patients?

The ways of thinking regarding human and artefact (machines of various relevant kinds) have implications that can be further substantiated on the ontological level. Ways of thinking are underlying preconditions that specify theoretical models of the relationship of human and machine. In ontology, the ways of thinking set out in section 1 are referred to as monism, dualism (pluralism) as the negation of monism and dialectical emergentism as a negation of both monism and dualism/pluralism.

They are further categorized in terms of human–machine ontology below.

3.1 Monistic Man–Machine Models: Technomorphic and Anthropomorphic Conflation of Concepts

Universalistic design is based upon monistic models. The human/society and machine are reified as one and the same entity, assuming the same level of complexity. They are deemed identical. This identity can be established in two ways according to the technomorphic and anthropomorphic praxiological conflations in section 1 (see Table 13.4).

Table 13.4 Monistic man–machine models

	"Man-machine" models		
Conflation	**Monism:** man/society and mechanism are identical, inasmuch as they share the same degree of complexity	**Reduction**	**Technomorphism:** any man/society is as complex as a mechanism
		Projection	**Anthropomorphism:** any mechanism is as complex as something human/social

3.1.1 Technomorphic Monism: Man—A Machine

Technomorphic ontology is reductionist: any human/society is a mechanism (machine). Both are mechanisms; both share the essential features of technology. By "mechanism" I mean an arrangement of entities or events that follows strictly deterministic connections. Strict determinism entails that given a determinate cause a determinate effect follows. Rafael Capurro (2012) considers this in terms of "patiens" and "agens". Strict determinism refers to "patiens" (patients). However, matter, nature, the cosmos is made up of self-organizing systems that show emergent and contingent properties and build up complexity in the course of evolution. Such systems follow a less-than-strict determinism. A determinate cause may entail different effects, and thus involves necessary but not always sufficient conditions. Here, we have "agens" (agents). To speak of a "mechanism" requires a shrinking of the space of possibilities that cancels out all possibilities but one, such that a single option remains for a system to take. Mechanisms are not complex. They may be complicated but cannot build up complexity by themselves only.

Technology as constructed by humans shall always yield a determinate output given a determinate input. Therefore, mechanisms are built that restrict the possibility spaces to one possibility only. Societies and humans are essentially self-organizing systems that include mechanisms only in subordinate positions.

Technomorphic monism is a fallacy. This fallacy is carried out by a concatenation of reductions following a series of steps:

1. The societal system is reduced to the individual actor, a fallacy of horizontal reduction of complexity (from the system to its elements)[5];
2. The individual actor as a social being is reduced to the human body as living system, a fallacy of biologism, which is a vertical reduction from social complexity (on a higher level) to a mere biotic complexity (on a lower level).
3. The human body is reduced to its physical substrate, a fallacy of physicalism, of reduction from biotic complexity to mere physical complexity.

[5] A reduction that has serious consequences as it eliminates social relations as the most essential part of the definiens of what is human. However, it is not because we enjoy the availability of language that we are relational beings but rather the relations that demand from us language-ability (see my Triple-c model of cognition, communication and cooperation in Hofkirchner (2013).

4. The physical substrate of the human body is reduced to a mechanism, a fallacy of strict determinism, of reduction from self-organizing systems at all to entities that have no capacity to self-organize.

Such a model of the ontic essence of man is used to substantiate a corresponding design: because man is a machine, it makes sense to treat man like a machine.

3.1.2 Anthropomorphic Monism: Machine, Human/Social in the Proper Sense

An anthropomorphic approach reverses technomorphic monism to achieve an identity. Mechanisms (machines) are said to essentially share human/social forms. This implies that the term "mechanism" must now be congruent with self-organization; it must include emergence and contingency. This is the case in info-computationalism (Dodig-Crnkovic 2014). The world is conceived as a computer, albeit not of the Turing type (input–output, etc.). According to info-computationalism, nature is always and everywhere engaged in informational processing ("natural computing").

An anthropomorphic approach commits a projectionist fallacy, similar to pan-idealism and pan-psychism. The fallacy follows a series of steps by which the complexity of a higher level is projected onto the complexity of a lower level:

1. The essential features of the social system are projected onto the individual actor.
2. The essential features of the individual actor as a social being are projected onto the human body as living system.
3. The essential features of the human body are projected onto its physical substrate.
4. The essential features of the physical substrate of the human body are projected onto *any* mechanism, be it natural or artificial.

The resultant anthropomorphic (projectionist) monistic model provides a foundation for the corresponding approach at the praxiological level: the artificial can be treated as human, because, in principle, it does not differ from the human.

3.2 Dualistic Man–Machine Models: Anthropocentric, Technocentric and Interactivistic Distinctions of Concepts

Monism requires an ontic unity of man and machine. Dualism, meanwhile, is disjunctionist and postulates an ontic duality of man and machine. This also involves a fallacy. Human/society, on the one hand, and machines, on the other, are viewed as separate entities, as different without anything in common. This ontic separation can, following section 1, appear as anthropocentric, technocentric or interactivistic (see Table 13.5).

Table 13.5 Dualistic man–machine models

	"Man–machine" models	
Disjunction	**Dualism:** man/society and mechanisms are genuine entities of different complexity	**Anthropocentrism:** man/society are of exceptional complexity
		Technocentrism: a mechanism can be higher complex than current man/society
		Man–machine interactivism: man/society and mechanisms are of different degrees of complexity but interact as if of the same degree of complexity

3.2.1 Anthropocentric Dualism: Man—Not a Machine

Human/society is modelled as something completely different from a mechanism (machine). For example, according to the Anti-Transhumanist Manifesto, first published in German in *Neue Zürcher Zeitung* (Spiekermann et al. 2017a, b), humans are:

> "animals of meaning", "enchanted beings", "capable of distinguishing between the state of being aware (mental presence) and the contents of which we are (intentionally) aware", using "attuned thinking and acting" (in original bold—W.H.), influencing "the emergence of our environment; thereby being co-creators of everything that exists", using "emotionality [in original bold—W.H.] as a basic principle of self-regulation and self- orientation", "attracted to the good", and "vulnerable beings" (2017b, 3-4); in a word, they are "sentient"; whereas AI can never be intelligent in a human sense (2017b, 2-3).

This might be read as epitomizing ontological thinking in social sciences and humanities. It expresses something akin to a theological position that does not acknowledge humans and society as product of evolution and increasing complexity of self-organizing systems. The assumed ontic uniqueness of the human provides a basis for exclusive practice and exceptionalism of the human "race". The human is deemed the pride of creation and is not in need of some other worldly goal beyond improvements by purely social (cultural, "spiritual") improvisations.

3.2.2 Technocentric Dualism: Machine—Superior to Man

Homo sapiens is, in principle, error-prone, whereas a mechanism (machine) is liable to failure only in the case of (a) operator errors, (b) programming errors or (c) material defects, all of which can be considered as "human failure". A technocentric dualist approach may then advocate "autonomous systems" that (a) work automatically, and without human interference, (b) program themselves and (c) build, service, repair and recycle themselves. The resulting concept of autonomy is one where a mechanism is modelled as something completely different from man/society. Autonomy, previously reserved for the designation of human freedom, becomes a purported property of artefacts. That autonomy, furthermore, would eventually limit human freedom and human autonomy.

Human obsolescence also becomes a possibility. Technological development has proven that mechanisms exist that can outperform human abilities in certain parameters, and these are taken as an empirical basis for the generalization that Übermensch that outperforms *Homo sapiens* in every parameter is possible and can be a goal of technological development. From an autonomous systems perspective, the qualitative leap from *Homo sapiens* to Übermensch—the singularity—will be set in motion by technology itself.

However, this qualitative leap seems doubtful. It seems to require the achievement of some agency on the part of Übermensch, implying that the successive achievement of technological states in mechanisms will result in emergence that demonstrates agency. This is mystic but not realistic. Agency is a property of self-organizing systems for which less-than-strict determinism holds, but not of mechanisms for which strict determinism holds. Emergence is the realization of a new possibility, but one that is grounded in reality rather than in mere hopes. Thus, human agency is arguably in a position to manifest qualities computers still cannot and will perhaps never be able to do (Braga and Logan 2017): purpose, objectives, goals, telos and caring; intuition; imagination; humour; emotions; curiosity; creativity and aesthetics; and values and morality. All these features make up a sense of self that has developed during millions of years of natural and, finally, social evolution.[6]

Anyway, the technocentric model of machines that can purportedly show higher complexity than man provides the rationale for the posthumanist design in favour of machines.

3.2.3 Interactivistic Dualism: "Actants" and "Intra-Action"

In interactivistic dualism, the human/society and mechanisms are modelled completely differently but as able to interact. ANT illustrates this position. Humans and artefacts are modelled as "actants". Both participate equally in a network in which they are assembled. An actant is something/somebody that/who acts and interacts and each is as causative as any other actant.[7] Consequently, the capability of social actors to control technology when producing or using it is conflated with the affordance of artificial devices. Of course, human actors under certain circumstances can behave mechanically, and whether humans turn out machinic or machines turn out human remains ontologically open. But this does not justify the conception that things can "act". For example, it is a misuse of language to state that the pistol shoots with me. It remains my person who shoots—with the pistol.

[6]See Porpora (2019) where he discusses thou-ness and personhood that he attributes to humans but also to non-humans, which makes sense if natural evolution developed those features. For that ontological reason, it would not make sense to attribute them to robots and other artificial constructs.

[7]See Donati (2019, 66 and 71).

Table 13.6 Emergentist man–machine model

	"Man–machine" models	
Integration	**Dialectic:** man/society and mechanisms are evolutionary products of nested complexities	**Technosocial systemism:** technosocial systems are emergent from man/society when mechanisms are functionalized

Sociomaterialism seems to go no further. Barad (2012) speaks of "intra-action" of agents as kind of activities of the world, of the dynamics of becoming, of a process of re-configuration, in which, however, there is no primacy—not of discursive practices nor of materialized phenomena.

In contradistinction to monistic ontology, in interactivistic dualism both human and machine are separated and identifiable, but ANT as well as sociomaterial interactionism provides a flat ontology. No emergent level that makes sense of real difference while allowing for synergy is considered. Instead, this interactionism switches back and forth on a continuum spanned between the poles of humans and artefacts. Though both humans and artefacts are conceded different degrees of complexity, they are conceived to interact as if of the same degree of complexity such that design practice is entitled to treat them both on equal terms.

3.3 Emergentist Man–Machine Model: Technosocial System Integration of Actor and Artefact Concepts for the Sake of the Whole

A dialectical relationship goes beyond duality. An integrative way of thinking in ontology is dialectical and emergent. It produces a unity of identity and difference. Unlike dualism, sides or parts are not completely separated. They are not brought together by an external operation, and they do not fuse completely in a flat ontology as is the case in interactivistic dualism. Instead, there are real differences and commonalities, and relations between the two, and these may be asymmetric. Emergentism focuses on the genesis of the relationship and the hierarchical ordering that is created (see Table 13.6).

Humans/society and mechanisms (machines) form part of technosocial systems. They have genetic ties, are evolutionary products and are encapsulated in technosocial systems. Technosocial systems relate them, insofar as systems harness mechanisms to serve humans/society. Technosocial systems emerge with the functionalization of cause–effect relationships for societal goals through technologies.

Social systems consist of actors (social agents) that are related by social relations (see Donati 2019). Actors can be individual or corporate. A corporate actor is not an individual social agent, but a collective one. That is, it is itself a social system, which is made up of individual—or other corporate—actors in certain social relations. In turn, as an actor, this becomes part of another social system—a suprasystem. A social system is a technosocial system, if the social functions of that system (that shall serve certain goals) are mediated by artefacts that are constructed as mechanisms. Arguably, artificial intelligence is and will be a mediation of collective intelligence of actors but is not and will never be an actor itself. What is labelled artificial intelligence is nothing that can become independent achieving a life of its own. Rather, it promotes the intelligence of a social system. In this vein, Francis Heylighen (2015, 2016) rejects the idea of a singularity by which a single suprahuman artificial intelligence would be possible, since intelligence is distributed over social actors connected by cyber-technology, such that the emergence of a "global brain" that remains rooted in humans is a more realistic scenario. From this dialectical point of view, what is in *statu nascendi* is a social suprasystem that is global, one that would represent a metamorphosis of humanity. Here, transformation changes the social relationships of society and not the technical infrastructure (alone). The ontic difference between actors and artefacts can be fixed in social, biotic and physical terms.

The ontic difference in social terms is that humans emerged from a change in cooperation of animal ancestors,[8] while machines are constructed by humans. Humans constitute social agency, that is action, interaction and co-action with other actors, which reproduces and transforms the social structure, and social relations that, in turn, enable and constrain social agency. Machines by contrast only support social agency and are not directly causative. Humans are the agents of social self-organization that produce and provide commons as synergy effects. Machines only pertain to the commons. Humans are the driving force behind social evolution, including the evolution of culture, polity, economy, ecology and technology, while machines are driven by social evolution. Humans can attempt to set off the transition which realizes a choice out of the space of possibilities, and that space is emergent. Machines do not trigger emergence themselves. Humans can reflect upon social relations and act accordingly, while machines cannot do this.

The ontic difference in biotic terms is that humans are autonomous agents that are able to maintain their organizational relations by the active provision of free energy, while machines are heteronomous mechanisms that cannot maintain themselves. Humans can make choices according to their embodiment, their embedding in the natural environment and in the context of their conspecifics, while machines cannot choose. Humans can try to control other systems, while machines are ultimately under the control of organisms; the complexity of the former encompasses the latter quite differently.

The ontic difference in physical terms is that humans represent an "agens" (agent) that is able to organize itself, that is, build up its own order using free energy and dissipating used-up energy, while machines represent a "patiens" (patient) that

[8]Which means a change in the kind of relations such that humans can no longer be defined without social relations as soon as humans have become part of a social system. As Donati says, "the social relation has its own reality, which is an 'emergent' not automatically derived from the qualities and dispositions of the individuals in relation" (2019, 61). It is co-emergent with the system that transcends the level of the interacting individuals.

cannot organize itself. Humans are made up of elements that produce organizational relations that provide synergy effects and can take part in suprasystems, while machines are made up of modules that are connected in mechanical ways. Humans work on the basis of less-than-strict determinism, yielding emergence and contingency, while machines are strictly deterministic, and neither emergent nor contingent.

I would argue that this ontology, including these key differences, is a cornerstone of any appropriate approach to human and artificial relations.

4 Epistemology: Cross- and/or Mono- or Multi- and Interdisciplinarity?

If we continue to follow a praxio-onto-epistemological approach, then our next point is to suggest that human–machine models depend on methodological frames by which the ontic relationship between human/society and mechanisms is investigated. Again, human–machine epistemology classifies research approaches according to ways of thinking. One class of frames prepares the ground for the corresponding class of models that prepares the ground for the corresponding class of designs. The frame classes are cross-disciplinarity, its negation which is mono- and multidisciplinarity/interdisciplinarity and the negation of all of them which is transdisciplinarity.

4.1 Cross-Disciplinary Man-Machine Frames: Technomorphic and Anthropomorphic Conflation of Methods

Monistic models use as their epistemological basis cross-disciplinary frames. Cross-disciplinarity means here, actually, that the methodology used cuts across different disciplines, while not recognizing the borders. It is taken for granted that the methodology for inquiring into social phenomena and the methodology for inquiring into mechanical phenomena belong to one and the same discipline. In this way, different disciplines are merged into one discipline. If we continue to use the terminology set out in sections 1 and 2, we can categorize two ways of crossing disciplines according to which discipline is the starting point and which is the target point (see Table 13.7).

Table 13.7 Cross-disciplinary man–machine frames

	"Man–machine" frames		
Conflation	**Cross-disciplinarity:** social data and mechanical data need identical frames for investigation	**Reduction**	**Technomorphism:** mechanistic frames are sufficient for social data gathering and analysis
		Projection	**Anthropomorphism:** social frames are sufficient for mechanical data gathering and analysis

4.1.1 Technomorphic Cross-Disciplinarity: Engineerability of Social Subject Matter

The reductionist position holds that a mechanical frame suffices for social inquiries. Social science is reduced to engineering science. Social phenomena are deemed "engineerable" by among others operation research, cybernetics, robotics, mechatronics, the fields of artificial intelligence and of autonomous systems. The methodology in question looks for mere mechanical ties in social phenomena. That is, if you start with a frame made for mechanical phenomena and cut across social phenomena without accepting the role of non-mechanical connections, you will end up in the model with mechanical conceptualizations only.

4.1.2 Anthropomorphic Cross-Disciplinarity: Engineering as a Social-Scientific Enterprise

An anthropomorphic, projectionist position involves a conviction that engineering science is, in principle, a social science. Mechanical figures, data and facts are said to have a social shape. A social science approach would suffice also for engineering. That is, if you start with a frame made for society research and cross over technology without elaborations, your model will exemplify the assumption that any mechanism is virtually as complex as something human/social.

4.2 Mono- and Multi-Interdisciplinary Man–Machine Frames: Anthropocentric, Technocentric and Interactivist Distinctions of Methods

A disjunctionist approach states that either discipline is self-sufficient; each monopolizes its own frame (mono-disciplinarity), or that the disciplines meet and have a superficial contact; they are "hyphenated" without giving up their identity (multi-/interdisciplinarity) (see Table 13.8).

Table 13.8 Mono- and multi-/interdisciplinary man–machine frames

	"Man–machine" frames		
Disjunction	**Disciplinarity:** social data and mechanical data need frames of their own	**Mono-disciplinarity**	**Anthropocentrism:** social data need pure social frames
			Technocentrism: mechanical data need pure mechanistic frames
		Multi-, interdisciplinarity	**Man–machine interactivism:** social and mechanical data need a mix of particular frames

4.2.1 Anthropocentric Mono-Disciplinarity: Sociologism

Social phenomena require only social science methods. From this perspective, technological issues are given no attention. This supports the modelling idea that man/society is of exceptional complexity.

4.2.2 Technocentric Mono-Disciplinarity: Technologism

Mechanical phenomena require only mechanistic frames. No attention is given to social issues. The focus allows the concentration on artefacts that might be higher complex than current man/society.

Sociologism and technologism work as parallel approaches without longing for a bigger picture. By their narrow frames they justify the supremacy of their own ontological model.

4.2.3 Interactivist Multi- and Interdisciplinarity: Socio-Technology

As a further permutation, segregation of the disciplines might also promote the juxtaposition of social science and engineering science. They co-exist, each is as valuable as the other, and there are no grounds for attributing supremacy in terms of what counts as social forms or what counts as technological forms. Both are conceived as reciprocally exclusive (Hofkirchner 2017a). Indifference is the result. This is the state of multidisciplinarity, a rather undeveloped state of working together. A further step occurs, when there is a cursory exchange at points of intersection, but without significant change or influence for either. This is a state of interdisciplinarity: a sociology of technology, an engineering of society, etc., none of which is in the position to overcome the flaws of one-sidedness.

A mix of frames, part of which collects and enquires social data, while the other part does the same regarding mechanical data, does not yield a satisfactory solution. What it does, instead, is cementing the indifference of the ontological models.

4.3 *Transdisciplinary Man–Machine Frame: Technosocial Systems Integration of Methods for the Bigger Picture*

An appropriately integrated approach requires a transdisciplinary frame. That is, "disciplines shall be transcended by the inclusion of a common code that shall perform the translation of concepts of one domain to those of other domains. By doing so, methodological knowledge shall orient towards the identification of similarities across domains to gain a deeper understanding of complex problems" (Hofkirchner 2017a, 2). Social data, mechanical data and data of their interaction need to be put together in order to yield a theory that helps us to understand the "big picture" (see Table 13.9).

True transdisciplinarity requires an approach that assumes an interrelation of disciplines in a systemic framework that grants relative autonomy to each according to their place in the overall framework; both social science and engineering should complement each other constituting a greater whole. That greater whole is achieved by shaping both disciplines in a systems perspective, that is, by viewing them as being part of systems science.

A systems science methodology has the potential to give the whole edifice of sciences a new shape—from philosophy on the uppermost level, via the triad of formal sciences, real-world sciences and applied sciences on the next lower level, to subgroups of disciplines of the former over disciplines to subdisciplines and so forth on ever more specific levels. A systems science methodology does not erect silos but assumes semipermeable boundaries and upward and downward interactions across all levels. If real-world sciences turn into sciences of real-world systems—material, living and social systems—and applied sciences into sciences of artificial design of those systems, the foundation is laid for the conceptualization of technosocial systems. Social science and engineering meet to construe a common understanding of the systemic relationship of society and technology. Each apply systems methodologies: for the development of technologies and for the empirical study of social systems in which:

> They can form a never-ending cycle, in which each of them has a determinate place: social systems science can inform engineering systems science by providing facts about social functions in the social system that might be supported with technological means; engineering systems science can provide technological options that fit the social functions in the envisaged techno-social system; social systems science can, in turn, investigate the social impact of the applied technological option in the techno-social system and provide facts about the working of technology (Hofkirchner 2017a, 7).

Table 13.9 Transdisciplinary man–machine frame

	"Man–machine" frames	
Integration	**Transdisciplinarity:** social and mechanical data need a single frame, comprising both on a meta-level	**Technosocial systemism:** social data, mechanical data and data of the social-mechanical interaction are put together in order to yield a theory that helps understand the bigger picture

Here, the epistemology of technosocial systems research paves the way for an ontology of human and machine, and for a praxiology of an integrated technology assessment and technology design cycle.

5 Conclusion

When imagining cyber futures, there are many fallacies that evoke in Anders's terms, hubris and humiliation, regarding social and technological practice (design), supported by one-sided theoretical concepts of agents and patients (models), which, in turn, are supported by methodologies of siloed disciplinarities (frames).

In order to avoid being trapped between logics that fuel pro- or anti-humanistic positions, an alternative logic is required that is, praxiologically, synergistic; ontologically, emergentist; and, epistemologically, transdisciplinary. Propagated deficient future imaginaries turn out to be not only undesirable but also unrealistic and, finally, unsound. In this chapter, I have set out the different categories in which this is the case. The consequence is a strong rejection of techno-optimistic and techno-pessimistic approaches, which obscure what is really possible, due to inappropriate justifications. The time has come to build momentum for an approach that shapes the future consciously and conscientiously. The proposed technosocial systems perspective is a cornerstone of such an approach.

Acknowledgements I am thankful to Jamie Morgan for criticism and polishing the English style in my drafts. The final version, however, is my responsibility.

References

Anders, G.: Die Antiquiertheit des Menschen, vol. 1: Über die Seele im Zeitalter der zweiten industriellen Revolution. Beck, München (1956)

Anders, G.: On promethean shame. In: Müller, C.J. (ed.) Prometheanism: Technology, Digital Culture and Human Obsolescence, pp. 29–95. Rowman & Littlefield, London (2016)

Barad, K.: Meeting the Universe Half-Way: Quantum Physics and the Entanglement of Matter and Meaning. Duke University Press, Durham (2007)

Barad, K.: Agentieller Realismus. Über die Bedeutung materiell-diskursiver Praktiken. Suhrkamp, Berlin (2012)

Bernal, J.D.: Science in History, vol. 4. MIT Press, Cambridge (1986)

Braga, A., Logan, R.K.: The emperor of strong AI has no clothes: limits to artificial intelligence. In: AI and the Singularity: A Fallacy or a Great Opportunity, Information, Special Issue. https://www.mdpi.com/2078-2489/8/4/156 (2017)

Capurro, R.: Toward a comparative theory of agents. AI & Soc. **27**(4), 479–488 (2012)

Committee on Legal Affairs. Report with recommendations to the Commission on Civil Law Rules on Robotics (2015/2103(INL)). http://www.europarl.europa.eu/sides/getDoc.do?pubRef=-//EP//NONSGML+REPORT+A8-2017-0005+0+DOC+PDF+V0//EN. (2017)

Corning, P.: The Synergism Hypothesis. McGraw-Hill, New York (1983)

Dodig-Crnkovic, G.: Info-computationalism and philosophical aspects of research in information sciences. In: Hagengruber, R. (ed.) Philosophy, Computing and Information Science, pp. 201–212. Pickering & Chatto, London (2014)
Donati, P.: Transcending the human. In: Al-Amoudi, I., Morgan, J. (eds.) Realist Responses to Post-Human Society: Ex Machina, pp. 53–81. Routledge, London and New York (2019)
Fleissner, P., Hofkirchner, W.: The making of the information society: driving forces, "Leitbilder" and the imperative for survival. Biosystems. **46**, 201–207 (1998)
Harari, Y.N.: Homo Deus: A Brief History of Tomorrow. Harvill Secker, London (2016)
Heylighen, F.: Return to Eden? Promises and perils on the road to a global superintelligence. In: Goertzel, B., Goertzel, T. (eds.) The End of the Beginning: Life, Society and Economy on the Brink of the Singularity, pp. 243–306. Humanity + Press, Los Angeles (2015)
Heylighen, F.: A brain in a vat cannot break out: why the singularity must be extended, embedded and embodied. In: Awret, U. (ed.) The Singularity: Could Artificial Intelligence Really Out-Think Us (and Would We Want It To)? vol. 19, pp. 126–142. Andrews UK Limited, Luton (2016)
Hofkirchner, W.: Emergent Information: A Unified Theory of Information Framework. World Scientific, Singapore (2013)
Hofkirchner, W.: Transdisciplinarity needs systemism. Systems. **5**, 15, 1–15,11 (2017a). https://doi.org/10.3390/systems5010015
Hofkirchner, W.: Creating common good: the global sustainable information society as the good society. In: Archer, M.S. (ed.) Morphogenesis and Human Flourishing, pp. 277–296. Springer, Dordrecht (2017b)
Latour, B.: Reassembling the Social: An Introduction to Actor-Network-Theory. Oxford University Press, Oxford (2006)
Müller, C.J.: Prometheanism: Technology, Digital Culture and Human Obsolescence. Rowman & Littlefield, London (2016)
Nora, S., Minc, A.: L'informatisation de la société: rapport à M. le Président de la République. Documentation Française, Paris (1978)
Porpora, D.V.: Vulcans, Klingons, and humans. What does humanism encompass? In: Al-Amoudi, I., Morgan, J. (eds.) Realist Responses to Post-Human Society: Ex Machina, pp. 33–52. Routledge, London and New York (2019)
Spiekermann, S., Hampson. P., Ess, C., Hoff, J., Coeckelbergh, M., Franck, G.: Wider den Transhumanismus: Die gefährliche Utopie der Selbstoptimierung. In: NZZ, June 19, 2017. https://www.nzz.ch/meinung/kommentare/die-gefaehrliche-utopie-der-selbstoptimierung-wider-den-transhumanismus-ld.1301315 (2017a)
Spiekermann, S., Hampson. P., Ess, C., Hoff, J., Coeckelbergh, M., Franck, G.: The Ghost of Transhumanism and the Sentience of Existence. http://privacysurgeon.org/blog/wp-content/uploads/2017/07/Human-manifesto_26_short-1.pdf (2017b)
Suchman, L.: Human-Machine Reconfigurations: Plans and Situated Actions. Cambridge University Press, Cambridge (2007)
Stanzl, E.: Maschinen im moralischen Dilemma. Wiener Zeitung, July 6, 2017. http://www.wienerzeitung.at/themen_channel/wissen/mensch/903014_Maschinen-im-moralischen-Dilemma.html (2017)
Trappl, R. (ed.): A Construction Manual for Robots' Ethical Systems. Springer, Cham (2015)
Wagner, P.: Techno Sapiens: Die Zukunft der Spezies Mensch. Documentation, broadcast on November 16, 2016, 3sat. https://www.3sat.de/wissen/wissenschaftsdoku/techno-sapiens-die-zukunft-der-spezies-mensch-100.html (2016)

Chapter 14
Where from and Where to: Transhumanistic and Posthumanistic Phantasms: Antichrist, Headbirth and the Feminist Cyborg

Britta Schinzel

Abstract Imaginations and desires of self-creation and infinite life have been known ever since humans are conscious about their knowledge about death. Such ideas are found in philosophies, literature or religious beliefs, and they are picked up and driven forward by today's notions, technologies and scientific investigations in transhumanism. In this article, questions of the genesis of such ideas are investigated, as well as the most different actual occurrences between biological survival and computational stamping detached from human physical life. The text will move between their feasibility and their desirability. After an introduction, the second section deals with the historical figures of infinite life, as there are golem, homunculus, the Russian movement of cosmism or the Czech invented concept of robots and their religious meanings. In fact, a deeper analysis of the transhumanistic ideas reveals a close linkage to the seemingly agnostic technocratic movement, or even a new technological religious meaning. In the third section, some less well-known stories of prolonged or infinite life in literature and opera are presented, showing both expectable economic and population developmental consequences, but also possible boredom of repeated experiences and other frictions. The fourth and last section will discuss some reflections stemming from Science Technology Studies (STS) and in particular from gender studies. The feminist concept of posthumanism criticizing some humanistic ideas is turned constructive with the feminist concept of cyborgism. And the very revealing discussion of gender in elite sports shows again how to elude gender dualism. This all leads to an end about posthumanism and the materialist theory of agential realism as argument and contention with transhumanism.

B. Schinzel (✉)
Computer Science and Social Research, University of Freiburg, Freiburg, Germany
e-mail: schinzel@modell.iig.uni-freiburg.de

W. Hofkirchner, H.-J. Kreowski (eds.), *Transhumanism: The Proper Guide to a Posthuman Condition or a Dangerous Idea?*, Cognitive Technologies,
https://doi.org/10.1007/978-3-030-56546-6_14

1 Introduction

Transhumanism is a biotechnically unrealized movement and a philosophy, also a "new religion of the technical elites", whose bio- and info-technological development is highly funded by the IT industry. Therefore, it will be interesting to follow the developments, also the unexpected ones, even if the prospective of its very realization might seem highly improbable.

Transhumanistic technology has predecessors in ancient human remedies such as eyeglasses or canes, prostheses and drugs. Today, however, these remedies are no longer just for body repair, but more and more for body optimization. Thus, medications are used to increase attention or mood, neuro-technological methods or exoskeletons to expand and improve the supposedly deficient nature of man.

Transhumanism is not a clearly defined term or even such a field of research, but an omnium-gatherum of ancient dreams and utopias of man, of material pharmaceutical, biological and technical aids, as well as literary, philosophical and scientific statements. Already age-old expansion possibilities of humans by material aids, like glasses or walking sticks, by drugs and medicines, fall into this range. Transhumanist or posthumanist visions of the future range in very different areas from biotechnology, invasive and non-invasive medicine and pharmacy, information technology, big data assisting methods of "artificial intelligence" and "machine learning", to brain emulation, i.e. uploading of the "inside of a brain" or of all brains on hard disk. Initiated by biomedical repair technology as well as the tools for elite sport, the technology on the body is changing from therapy to improving and optimizing the self. The technology provides artificially produced organic as well as non-organic spare parts, which can be manufactured with 3D printers, for example. In the scientifically manipulated degradation and structure of the body, the border between "nutritional supplement" and "medical treatment" and between therapy on the body and artificial improvement of the body is blurred. Newest developments result from the ever-closer connection between sciences and technology in the NBIC (nano-bio-info-cogno) technologies and sciences, also forming all the duplicated subjects with the prefix "computational-", with artificial life research, trans- and posthumanism and exproprianism. Traditional demarcations between man and machine become fuzzy, such between the natural and the artificial, organism and artefact, the grown and the produced.

Who can determine what is ethically and morally permitted in this transformation? The line between the forbidden and the permitted becomes increasingly negotiable. It is moved into an interactive process where moral and social rules are adapted or adjusted. Defining these borders and taking control of border crossings has largely been left to industry and individual scientists and is only commissioned ex post in national and international health and ethics committees.

Finally, transhumanism further promises the immortality of individual identity as an even medium-term goal. The prefix "trans-" signifies a transition to posthumanism, which is supposed to leave human existence behind as an evolutionary step.

In fact, it is not a unified movement, but a wealth of different methods, techniques and fantasies, from biotechnological interventions to downloads of human brains and brain activity, with the ultimate goal of being able to leave the earthly bound existence and completely into virtuality to enter. As developments are driven forward by the rich funding of a research institute by Google, critical questions are posed through science technology studies (STS), technology impact research (TA) and gender studies (GS).

Antonio Gramsci's dictum on the optimism of action and the pessimism of knowledge can guide the field of tension between the fantasies of biotechnical making and the reflective evaluation. Here, I will only mention a few utopias and notions preparing for the ideas of self-creation and infinite life, as are golem, homunculus, cosmism and robots. The text will move between questions of the genesis of such ideas, their feasibility and desirability and in particular some ideas stemming from literature and reflections from gender studies.

2 The History of Ideas and a New Religion

In all religions, the gods and goddesses are living forever, also the Greek ones, and they are able to turn mortals into immortals. Kalypso, e.g., offers Odysseus to live forever, but he renounces this; he wants to and must go back to Ithaca. Greek thinking also lends ideas of virtuality and infinite life of ordinary humans to transhumanism. Plato, with his allegory of the cave, understands man, after his ascent from the troglodyte from the illusory world of the cave to ever higher levels of knowledge, as essentially immaterial, soulful, insofar as virtual, as he appears in the form of a body, but persists after its decline in the realm of ideas. Pythagoras imagined mathematics as the aid to advance into true reality, the immaterial, transcendental realm of numbers. As with Plato, for Pythagoras the immortal soul is trapped in the body, and only after death will it be able to return home into the kingdom of numbers.

But also the idea of man producing the foundations of one's own existence has old roots. As new, forward-looking and future-oriented as this "religion of the technical elites" may appear, it is based on old thought figures, phantasms and images of man. Ancient myths anticipate the creativeness of living beings through man.

For the one-God religions, this self-creativity is sometimes imagined as the original sin of man, in Jewish religion the false Messiah, the Jewish Armilus, in Christianity the Antichrist, in Islam the Dajjal. In the Jewish conception, according to the apocalyptic end-time expectation, the evil, which empowers itself in ever new systems of violence on earth, becomes independent and becomes personified as the counter-god. The Antichrist (Fig. 14.1) of the Original Christians from the secret revelation of the evangelist Johannes of Patmos is the inner and outer opponent of their faith, counter-power of the Creator-God, Jesus Christ, who is expected before his Second Coming. Jesus himself foretold its appearance (synoptic apocalypse Mk

Fig. 14.1 Antichrist window in Frankfurt (Oder); around 1367 Antichrist and his pendants; stained glass

13,21ff EU par.) "The Antichrist will gather followers around him, . . . and promise eternal life."

The idea of self-creation of man already appears in the Jewish folklore with the golem, originally created by Rabbi Eliyahu in the sixteenth century, or a little later, the Prague golem of Maharalas. It also appears in alchemy with the homunculus (e.g. Paracelsus' in "De natura rerum", 1537), this in various fantasies of its creation.

In the early nineteenth century, the "homo faber" appears in Goethe's Faust II laboratory scene, where the homunculus stands for the transgression of human nature and purification of reproduction through science and technology. And Mary Shelley created a dystopic figure with her monster Frankenstein. The thread continues to weave on the evolutionary human image since Darwin, where man is also regarded as "intermediate being". In fact, transhumanists often refer to Nietzsche's Zarathustra (Nietzsche 1885), with: "Man is something to be overcome." Man appears as a creature changeable, shapable. It was even more true at the beginning of the twentieth century, when biologistic eugenics was meant as socio-economic progress, even as an obligation of humans to take their evolution into their own hands. In his reflections on anthropotechnology, Peter Sloterdijk now takes up Nietzsche's so-called bridge existence, the "pontifical" existence of the human. Man as Homo Faber forms with his technical means not only the world, his outer frame of reference, but with it always also himself. Thus, he also changes the relation to himself, namely to owe himself to himself, by producing technologically the foundations of his own existence.

Russian scientists and artists in the late nineteenth and early twentieth centuries founded a movement called "cosmism", which pursued the goal of making people immortal through technology (Groys and Vidokle 2018). In this way, the Christian promise of resurrection should be redeemed in the here and now with the overcoming of mortality and with space colonization. They also experimented with blood exchange. Central to this endeavour however was the arts, which, because of their ability to foresee things, were considered as equal to science and medicine. They would be able to find forms of the creative development of the mind, and heaven and earth would form a unity with man harmoniously fitting into it. These ideas transferring art exhibition in NY in 1920 influenced also cosmopolitan artists and thinking in the USA. Cosmism, among other philosophical currents, also laid the basis for the twenty-first century thinking of new materialism and transhumanism.

The technology philosopher Oliver Müller sees a philosophical tradition of transhumanism in Sartre's programme of existentialism in his essay "Is Existentialism a Humanism?" of 1946, which he understands in his pathos of self-creation as the precursor philosophy of transhuman or posthumanism. Thus, the design character of (contingently imagined) human existence in existentialism appears to be reflected in the technical design of the transhuman. Because Sartre claims in his essay that the existence of the essence precedes: humans are the self-producing beings. Müller then investigates to what extent transhumanism uses implicitly existentialist motifs and thought figures and where it differs from it. If the biotechnical intensity of experience increases, then fear, despair and guilt should lose their existential power. The challenge of one's own death should no longer lead to a struggle for orientation, but to the abolition of this probably greatest provocation of being human. In this opinion, transhumanism could even be understood as an answer to existentialist heroisms. On the other hand, the transhumanist is clearly limited in his possibilities, because he can only choose such a design of his own, which can be produced biotechnologically.

Also other utopias of a better or infinite life, and self-replacement wishes of the homo faber, are all footing in the culture-historical mortifications of humans. The mortification of Galilei's decentring the earth, of depriving God's creation of mankind through Darwin's evolution theory, was turned into the subsequent self-gratification through eugenics, a nature replacing human, the work on him/herself into a designed plannable future.

The attempt to grasp the brain by computer simulation or download with the therefore necessary formalization and mathematization shows an apprehension of the human within a specific relation to the world. According to Th. W. Adorno (1959), humans want to bring the irrationality under control, as they experience nature as irrational and thus threatening. Therefore, it is to discover the rules

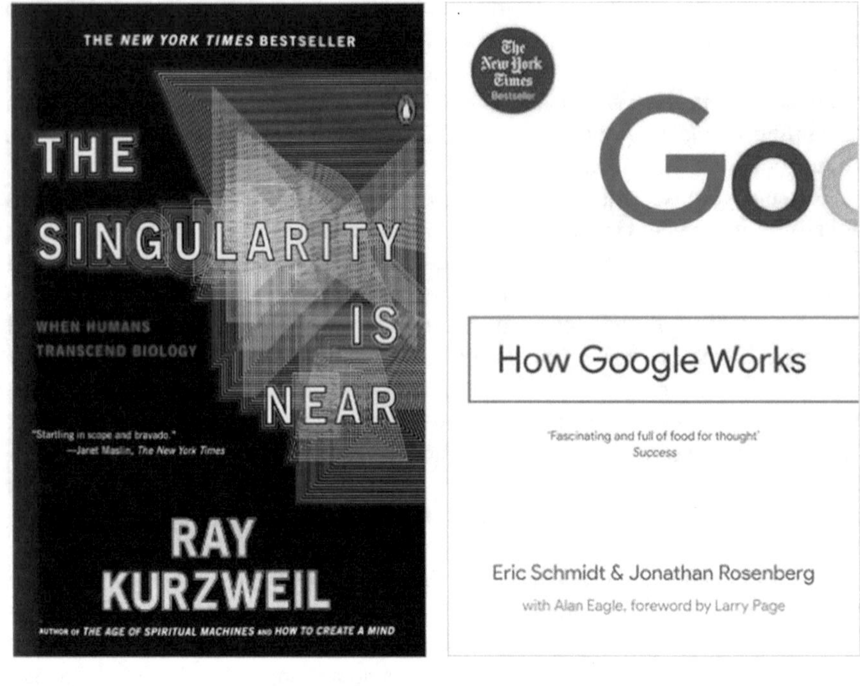

Fig. 14.2 The book covers of two related books

prevailing in nature, to cultivate them by means of technology and ultimately to control them. The technical penetration of nature is also directed at the nature of the human, her body and finally her and against her inner nature. It is precisely this inner nature, creativity, inventive capacity, but also feelings and affective expressions, which seem to elude control, in which irrationality predominates. Trying to control this area with the aid of technology, Adorno states that the root of technology itself is not technical, even not rational. He states that nature triumphs by virtue of its taming again over the tamer, who, the latter, resembles it by strict scientistic objectivity. In the process of such assimilation, the elimination of the subject for its self-preservation, the opposite of what he knows about himself, asserts itself, the mere inhuman relationship to nature. Human mind becomes obsolete in the face of progressive mastery of nature and is struck by the stigma of magic, which it once imprinted on nature: it foists subjective illusion instead of the force of facts.

Immortality is imagined in Google's science laboratory "Singularity Lab" by downloading the human brain in a future that is no longer the future of humans but of posthuman beings that once were humans (Fig. 14.2). The singularity event, when death is over, is expected by Google's chief engineer Ray Kurzweil for 2030. Oxford philosophy professor Nick Bostrom, the head of the "Future of Mankind Institute" at the local faculty of Philosophy and the James Martin 21st Century School, sees the year 2045 as the probable time when technology could develop something like "consciousness" or, as their jargon calls it, become a singularity. Google's project

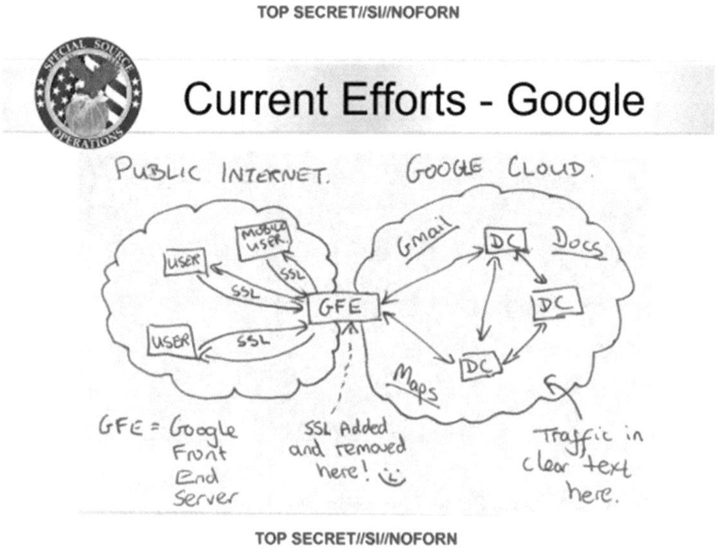

Fig. 14.3 NSA muscular Google cloud from Snowdon leaks

"End Aging and Death" seeks to make information intelligent by machine learning on big data, then to eradicate disease and multiply the lifespan of the human body, eventually defeating death. Individual technologies such as the prevention of telomere shortening or genetic modification are to be perfected and integrated into artificial intelligence for individuation. Then technology could "intelligently" affect human level. Zoltan Istvan, the US presidential candidate of the Transhumanist party, says that "we can worry about many other things later, like social equality, ending wars, and ending poverty. But first we must strive to create technologies that stop the world's greatest killers: aging and disease." Big-headed self-gratitude and the associated implications of the idea to be self-presupposition of oneself are positioning man upon the foundations of one's own existence.

But maybe instead of Google positioning (itself as) the new God in collecting all the world's knowledge in its Google Front End (see Fig. 14.3), this privilege might be awarded to the NSA having all the world's hardware wires, Internet layers, software and data under control (see Fig. 14.4), as Bernhard Taureck (2014) fantasizes. Anyone who has much more knowledge about people than they have about themselves is already in a superhuman position towards the populations.

The preeminent knowledge of the national institution gives people the feeling both of being nothing and of being embedded in the big data pool: NSA knows everything about him in detail and potentially forever. Being conscious of this superhuman informedness, this informational transcendence can lead to fear and horror in the individual. But he can also experience this dependency as a feeling of caring security and share this living with other people. In order to justify this superhuman knowledge monopoly, there seems to be no choice but to invent a

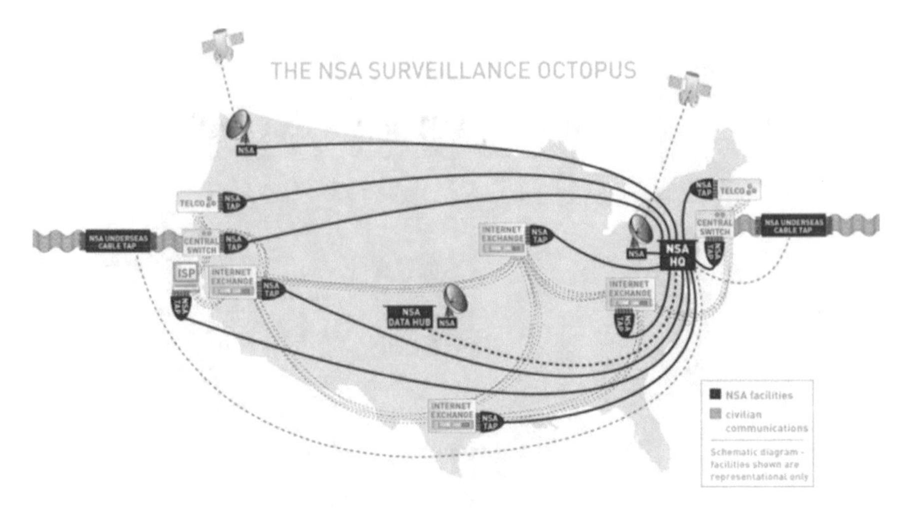

Fig. 14.4 The NSA surveillance Octopus from Snowdon leaks

religious vocabulary. And in fact, Taureck reveals that an ideologically upgraded NSA religion would fulfil the two conditions of religion: according to Schleiermacher, the consciousness of each one to hold little power and ultimately to depend on something much larger and more powerful than himself, and on the other hand, according to *Durkheim*, the beliefs of a religion are shared by communities and connect the believers. Both conditions seem to be fulfilled by an NSA religion. But moreover, the conviction in contrafactual fiction, especially religious ones, is welding communities together, and this seems so the more it is on the cost of sacrificing the intellect. This also holds true for transhumanism, but it has the rational vindication as well, often drawn from the humanistic tradition of elucidation that the possibility of an infinite life in some future cannot be disproven.

3 Transhumanism in Fiction

As the recent transhumanistic fiction in English language is widely documented (see, e.g., https://en.wikipedia.org/wiki/Transhumanism_in_fiction), I will only mention three pieces of fiction, one where an alchemist drug allows (finite) life prolongation repeatedly, one talking about the process of cryopreservation of living persons (in fact up to now no living mammal has survived such a process after unfreezing) and one reasoning about biological engineering of cell reproduction. The computer-supported brain upload and posthuman life in silico is not discussed in my examples. But I would also like to mention the series "Ad Vitam—Forever" in the French-German TV channel "Arte". There vicious conflicts are told, between people who want to be—and as considered "normal" even are forced to be—"regenerated" forever, and those wanting their life to end. Also a "Christian" religion arises violently enforcing the death of young people before the age of possible "regeneration". The films describe the—not only lying in the future—social pressure to genetic and physiological modification within the capitalist logic of accomplishment and success, leaving so-called losers as guilty for their fate behind.

In his play "Rossum's Universal Robots", Karel Čapek envisioned a robot revolution that would overthrow the dominant humans and lead to their extinction. With this the word robot (original meaning slave) was formed, furthermore meaning an automaton, a self-operating machine.

Even more interesting in our context is his play "The Makropulos Secret" (Věc Makropulos), which Leoš Janáček had adopted in a libretto for his opera "The Makropulos Affair", investigating the implications of prolongation of life up to human immortality. The protagonist with the initials E. M., as Emilia Marty, 330 years old, lives among people with normal lifetime, in different countries, under different names. Through all her lifetime she is a famous opera singer. She is mother, and multiple ancestor, i.e. mother of mother of mother,... of men, who were or are her lovers and bears in her memory all the circumstances of her life. The story describes all the entanglements stemming from the contemporaneousness of ages, and the boredom of a person having lived through all the joys and problems of existence, incidents, tensions of love, hate and misery.

She looks like about 30 years of age, but although most beautiful, there are some signs now of a strange kind of ageing, which bedazzles some of her admirers. More than 300 years ago, when she was 17 years old, her father, an alchemist, had tried a drug on her that would give her a 300-year survival as young person. As contemporary Emilia Marty she bursts into a lawsuit that has been conducted between two families, Prus and Gregor, for generations around the inheritance of a Baron Prus. Emilia miraculously knows a lot about the families of Gregor and Prus; she gives the information that the ancestor of Gregor is the son of a baron Prus and some Ellian McGregor. She hints on the location of an unknown testament, which might or should change the outcome of the process in favour of Gregor. As Emilia had centuries to train her voice she sings heavenly and in addition she is unearthly beautiful. The audience is at her feet and all men fall in a kind of deadly love with her, because she is abysmally fascinating. The old Count Hauk-Šendorf thinks he recognizes Emilia as Eugenia Montez, a Romani woman with whom he had an affair in Andalusia half a century before; she tells him Eugenia is not dead and again falls in love with him. She reveals that the mother of Baron Prus' child was Elina Makropulos and another identity of her was Ekaterina Myshkin. In order to get hold of an envelope which contains the prescription for preserving her youth for another 300 years, because her father's drug ceases working and she starts to become old again; she spends a night with Prus in order to get hold of the potion, in which she succeeds. But finally, Emilia realizes that perpetual youth has exhausted her and made her bored and mopish. And that sense of transcendence and meaning can only come from a naturally short span of life.[1] So she decides to allow her body a natural ageing up to death.

In order to envision the circumstances of cryopreservation[2] of a living person, the novel Zero K by Don DeLillo is of interest. It moves philosophical questions about human freedom to choose one's fate. The billionaire, Ross, whose wife Artis is deadly ill, wants to meet her in the future, when healing or a transcendent life is possible. Eventually he decides to let himself be frozen as well, until she might be cured, seeking immortality for both of them. Ross is investor in a secret compound where death is exquisitely controlled and the novel leads through the long-lasting and complicated process of biomedical and technological processes until she surrenders her body to the cold preservation until a future time when biomedical advances and new technologies can return them to a life of promise. Jeff, Ross' son and the book's narrator, is committed to living, to experiencing "the mingled astonishments of our time, here, on earth". He sees the long-lasting medical process of the cold-technological hibernation of his step-mother; he is confronted with the already deeply frozen bodies, often heads teared off for enabling more differentiated processes within the latent life, subject to their return to the usual course of time in an arbitrarily distant future.

Don DeLillo's novel spectacularly observes and weighs the dark sides of the world—terrorism, floods, fires, famine, plague—against the beauty and humanity of everyday life; love, awe, "the intimate touch of earth and sun"; and the contradictory yearning of the living for the cold death, and crying to be saved from death, hoping for healing after thawing, i.e. dead organismic matter hindered in decay. Humans who perhaps suffer an incurable deadly disease, which they hope would be treatable later, are put into a frozen undead state by means of refrigerators. Frozen, the living and the dead are approaching in a latent life. In suspended animation the processes of life are slowed down so that they resemble death, but only almost, because the ageing and decay processes are also suspended. Successfully thawed, the resurgent life resumes its self-preservation and ageing processes: suspended animation. It is a paradoxical clinging of the living to the dead, the cryogenic prolongation of life, a life that adapts itself to the dead human in order to escape death.

Thea Dorn's German novel *The Unfortunate* (Fig. 14.5) deals with the aiming at immortality through biomolecular engineering. Johanna Mawet is a molecular biologist, who researches in zebrafish, the cells of which have an infinite capacity to replicate, a chance for immortality. During a research stay in the USA, she meets a strange, ageless gentleman. The closer she gets to know him, the more abstruse are the experiences with him. His story, which he tells her, seems like stemming from mental illness: "I was famous! Herder, Humboldt, Schlegel, Arnim, Brentano, the

---

[1] A recent investigation in the desirability of an infinite life among American students in fact brought the result that they considered it to be probably very boring to live overtime.

[2] Francis Bacon already formulated the utopia to be able to stop the ageing by cooling in: New Atlantis, publisher William Raleigh, London 1669. See also the text Kryierung: Alexander Friedrich_ Outsmarting the transient; in Jahrbuch Technikphilosophie 2016 (Ed. Gerhard Gamm et al.): List and Death; diaphanes Zurich - Berlin.

Fig. 14.5 Thea Dorn's novel *The Unfortunate*

Duke of Gotha—all of them admired me! Even Goethe sought my advice!" He claims to be the physicist Johann Wilhelm Ritter, born in 1776. Not by chance Dorn choses the physicist Johann Wilhelm Ritter for her story, because he both had researched in electricity and speculated about the nature of Nature, both leading him to research in occultism. 240 years later, he was now suffering from his immortality; he was tired of life, but every suicide attempt fails. In order to track down his supposed immortality, Johanna lets his DNA be sequenced and investigates his body's ability to regenerate, until her findings create doubts in her former repudiation of his story of provenience. When Johanna's colleagues become suspicious, she and Ritter fly to Germany, where she tries to pursue her studies on his cell physiology. But in the end, she loses her job, and getting more and more involved with her test object she falls in love with him. Finally, together they find a way out by drowning themselves—it is not explained why now in contrast he succeeds in suicide.

The American narration puts the elements of reflection into the mouth of its protagonist, whereas in the European ones the story itself contains and reveals problems with artificially prolonged life, leaving the reflection to the reader. Summarizing the examples of fiction the American view on transhumanistic ideas seems to be positive, promising and hopeful, also reflecting ethical and philosophical questions, but under the supposition of a very prosperous society and infinitely available or reproducible resources. In contrast, the European one seems much more critical and hesitating, maybe considered to be retarded, backward-minded contemplators by the former.

4 Transhumanism, New Materialism, STS and Gender

The convergence of nano-bio-info-cogno (NBIC) technologies already has great potential for the technology-driven and irreversible transformation of man. Machines and humans are measured, monitored, analysed and subjected to technical control. Humans will more and more be infiltrated with technology, will assimilate to computers and will be exploited by them. For biotechnological use, human organs are already bred in animals for whatever use. How to determine what is ethically and morally permitted in this transformation? The line between the forbidden and the permitted becomes increasingly negotiable. Also, the critical voices of gender studies and science technology studies provide us with quite different answers.

A focus on gender does not only mean, and even least, that the gender issue is addressed in the context of the changes brought about by enhancement technologies, but that transhumanist ideas, developments and publications are analysed and evaluated from a gender-theoretical point of view.

Gender studies give attention to transhumanism from various scientific points of view, and using various theories, all intertwined with the technological ones. Relevant methods and positions of gender studies in the field of STEM and science and technology studies (STS) are, e.g., the (discourse) analysis of narratives presented in the respective context in scientific publications, in fiction and narrations, e.g. transhumanist science fiction; Judy Wajcman and Wendy Faulkner among others theoretically analysed the synchronous co-construction, even co-materialization of technology and gender; Lucy Suchman and Donna Haraway emphasized the juxtaposition of all technological research with practice, i.e. all design and placement and acquisition of technology; and Haraway stated her political programme, cyborgism, taking the human–machine entanglement positively and making it ready for change, leading to the gender theories of feminist materialism, such as Donna Haraway's agency and cyborg theories; more recently also Karen Barad's agential realism stemming from her philosophical analysis of quantum theories appears to be capable of dealing with transhumanist developments and ideas in trans- and posthumanities.

Since the beginning of the feminist theory of science in the 1960s, later leading to and partially merged in gender studies as a science, the programme of rational science initiated in Renaissance by Francis Bacon and Renè Descartes has been criticized as androcentric. This because the rational method of the natural sciences is based on separations and binary oppositions, in particular the subject–object separation. It positions the male researcher as the subject against the female-coded nature as an object, in order to "wrest her secrets" from her by means of Bacon's decontextualizing experiment. Humanism, born with the rational method as a philosophical-political-ethical concept, again places the rational man into the centre of creation. He is to subjugate the—symbolically female—earth, morally vaccinated by the Christian God. It is just the entanglement of humanism with racism, colonialism, Eurocentrism and androcentrism; why some feminist critique of the technosciences led to concepts of posthumanism. It is setting aside humanism and should not be confused with the posthuman ideas of transhumanism, leaving human beings to be overcome by evolution. But in fact the feminist posthuman also links up with transhumanist ideas and technical developments in such a way that humans and technology merge, the boundaries between humans and animals and other beings become permeable, new intermediate beings emerge, as vampires or elves already are: cyborgs.

4.1 Cyborgism

First, the cyborg, a combination of human and machine, had been imagined as a military concept to overcome space problems. This transhuman cybernetic organism, paradoxically created by man for autonomous autopoietic development, should be a self-governing being, beyond creationism and evolution. However, let it be imagined consciously or unconsciously; it was the androcentric dream of overcoming the deprivation of being born out of a woman, the headbirth. Taking the merging of living beings with machines as model for man–machine communication, Donna Haraway (1991) turned this concept into a feminist project with her cyborg manifesto. The property of the cyborg to dissolve the boundaries between humans and machines, but also between humans and animals, is a political possibility to eliminate hierarchies and dualisms. However, Haraway (2007) takes distance from the notion of transhumanism, of posthumans or posthumanism; she is not concerned with human improvement, but with decentring the human, dismantling the uniqueness of humans.

4.2 Inclusive Humanism Contra Optimization Ideology

We live in a techno-biological environment and are continuously delegating cognitive and physiological processes to whole networks of things. However, nothing prejudices that the mechanization of the body has to signify a vitalocentric increase in efficiency, a rise from repair to improvement, from therapy to enhancement. Gender researchers like Karin Harasser (2016) criticize the upward trend from repair to entrepreneurial self-improvement, optimization, performance and perceptual enhancement. The prevailing global growth ideology is driven by economic motives, and—wrongly—refers to Darwin's concept of evolution (Harasser 2016). But rereading Darwin's original texts on his theory of evolution she reveals that his concepts of adaptation and fitness are always relational and situation-specific and—

unlike as it is usually thought today—not pointing towards a uniform upward trend but are always related to a milieu. She argues that our imagination about the development of our body can only be founded in the past, whereas its future is open and contingent. Nor can we anticipate the future of man–machine relations, i.e. "the techno-body is to be understood as one of whom we have always known afterwards what it would have been capable of". She resumes, we should not project it negligently into the future of its perfection, and thus deliver it to the pressure of constant self-improvement. With such an understanding of the body she considers "being human" not as an immutable quality, and an inclusive humanism as a horizon that can potentially include many that are not commonly considered as human. It is aimed at the arena of action, which gives space to only partially sovereign actors (who, after all, we all are). Haraway uses the word response-ability for the idea of a political arena in which innumerable actors should have the ability to articulate, to be heard and to be able to contradict.

4.3 Elite Sports

Consumer industry already offers many portable devices to support and change body and mind. Among others Katharina Dermühl (2015) states that through gadgets, apps and sensor-based technology, our body is made into a quantifiable object, both by ourselves and by the healthcare industry. The goals of "the quantified self" (cf. Duttweiler et al. 2016) are to make mind and body accessible for improvement. Social conditions facilitate these transformations from "normal" to "improved", such that normality becomes a flexibly situated value, clearing the way for the acceptance of also invisible technologies, like genetic engineering.

In elite sport the perception as normal already is a power game between conservatism and change, and it is strongly gendered. Conservatism manifests itself, for example, in a strict observation of gender segregation. And the monsters of the sport, the bio-amazons, and the cyborgs are using change technology to push the boundaries drawn by conservative supporters of fairness ideology.

The area of elite sport sometimes is described as an already posthuman world, a space in which man himself is the object that is being redesigned. This led Cecile Crutzen (2003) to deeper investigation, asking about the concessions: who may decide juridically, organizationally and morally about development and application. She considers elite sports as a critically transformative space, where athletes have already "metamorphosed into cyborgs, kind of super-humans, ambassadors of transhumanism, placed at the cutting edge of human boundaries of capability" (Miah 2003). The design of the human body and its clothing take place both physically and mentally. "The athlete's body is in a state of flux, continuously transcending itself, and thus perpetuating transhuman ideas about the biophysics of humanity" (Miah 2003). "Elite sports is not just about running fast. It is about the interaction of medicine, diet, training practices, clothing and equipment manufacture, visualization and timekeeping" (Kunzru 1997).

The designed artificial-technoid human, however, does not leave her or his personality untouched: doubts about self-worth of one's own achievement, or the artificial one; fear of identity loss and fear of risks and health consequences are among other problems to be dealt with. The sports world is a critical transformative space where one can test the reliability—the spectrum between trust and discard (Crutzen 2003). Unfortunately, the athletes are to the least the real actors in this space, but the sponsoring industry, the sports institutions that determine which technological means are allowed and the media. The spectators, only passively present in this room feel attached to each, offered reflection of heroism by taking risks for surpassing limits.

If elite sport continues to pursue the goal of expanding the limits of the biological possibilities of man, it will discover that it cannot do without technological enhancement (Miah 2003). This means that the question of whether doping is or is not unnecessary is to be answered in regard to this goal.

With the excuse of fairness necessities, however there is an unfair and unethical testing of female athletes in elite sport. In the system of rules of the sports associations, the privacy of athletes is neglected in a way as it would not be allowed for a normal citizen. As a requirement for transgenders to attend Olympics, the IOC required these participants, as the sportswoman McArdle said 2008, to have undergone surgical anatomical changes, and in addition to having received appropriate hormone therapy in order to minimize gender-related advantages in sports competitions and to have had their change of sex recognized in their host jurisdiction.

Crutzen's proposal to overcome these discriminating practices is to allow athletes themselves, as users of the technology, to design their ideas of a changed person. The doping control industry may retain its expertise in the field but should change its function to "care rather than control" so that athletes can experiment in safe environments, where they can leave a design if they do not like it, and where they can decide to live consciously with any long-term consequences of an irreversible design. Having changed in this way, discrimination based on supposed gender characteristics will be unnecessary. Also, female athletes, the monsters of sports that transcend the arbitrary female–male border, would no longer be portrayed as the fraudsters who hurt the rules. Athletes, on the other hand, would be empowered to take ownership of their bodies, be designers of their own identities, releasing the enhancement technologies into their power of disposition, if they are fully aware of all the consequences and risks. In a change process they deconstruct the gender constructions commonly used in elite sport by negotiating the cultural norms of femininity. Moreover, according to Crutzen (2018), the sports arena could open up a larger experimental space for all. Gendercrossing in elite sports could spread further into society eliminating the boundaries between the genders, and even making disappear some differences between the sexes.

4.4 *Posthumanism, Agential Realism and Diffraction: How Does It Link to Transhumanism?*

Following Haraway, who considers technologies as materialized figurations (narrations giving not only meaning but also agency), as bidirectional material-semiotic processes, which define, stage, restrict, limit, control and being able to generate powerful social and material processes, Karen Barad (2007) introduces the terms "intra-action" and "agential cuts". While in interactions already formed entities are going into exchange, in contrast the concept of intra-actions reverts this, such that separated entities are the result of such material-semiotic processes. Human–machine interfaces, e.g., are the place of such intra-actions, where humans and machines are configurated simultaneously. The interface is not a given borderline, but an enacted one, the effect of which unrolls itself at use in the respective configuration. In science, the object of investigation, the instrument and the observer are forming a unit, the "apparatus". Agential cuts are denoting those material-semiotic actions of drawing borders, which are common efforts of language and material. All the reconfigurations undertaken by transhumanist enterprise are material-semiotic actions, think of science fiction literature, scientific figurations, their materializations and all the symbolic and material recursions driving the developments forward.

Karen Barad envisions a future, where machines will generate new life. As a quantum physicist she focuses on physical cyber systems on nanoscale. Cyber-physical systems (CPS) are integrated physical processes with computation and networking, and like claytronics (matter manipulated at the nanoscale) or nanotechnological smart dust (nanoscale particles packed with a power supply, sensors and a means of communication), they function invisibly. They inter- and intra-act with humans and non-humans in the background through many new modalities. Agential cuts and amalgamations of these systems have many impacts on the world and on humans, as it is on health or persistent surveillance. Life will be reworked through nanoscale life processes, and the combination of computational nanotechnology and bio-nanotechnology in neuro-electronic interfaces is joining computers to the human nervous system. All these immense changes will also alter our bodies, desires and imaginations and amend our propositions of "who we are".

In the following, the posthumanist positions of feminist scientists will be considered through the lenses of sex and gender, nonetheless holding in a wider frame. Posthumanism[3] is understood by Barad, using positions of new materialism, both as a critique of the anthropocentrism of humanism and of anti-humanism, by critically examining the borderline practices between what is considered human and anything else. She further differentiates posthumanism from subject positions that determine man either as a cause or as an effect, and as the body as a natural or solid dividing

[3]In this section, I will follow Waltraud Ernst: Menschliche und weniger menschliche Verbindungen; Posthumanismus und Gender; FIfF-Kommunikation 3/2016, pp. 37–41.

line. Difference is not assumed in this approach; rather their production processes are examined.

Barad considers matter, biological as well as hard one as fluid, as "intra-active performative", even when it comes to understanding sex. The concept of the apparatus serves to describe the processes of production and normalization, in which, e.g., two-level sexuality is produced by means of different conceptual fields. The term apparatus includes more than one lens through which one sees something or a construction. It points to a complex interaction between social norms and the physical states of people charged with meaning. In the performative display, e.g. gender relations are produced and lashed down.

With her approach of agential realism as a new feminist epistemology, Barad suggests understanding matter and materiality, including the body, as a dynamic, intra-active becoming. It also incorporates non-human organisms and non-organic matter, questions the sharp boundaries between organic matter and non-organic matter, as well as the clear boundary between the organism and the environment, and opens the space for a wide variety of natural systems of reproduction and mutual intra-active attachment. Here again the role of prosthetic aids comes into play, but now together with Barad's symbolic overtaking of the physical effect of diffraction. Diffraction patterns arise when a wave meets a barrier or a slit. They are used as a metaphor for making differences, the production of difference patterns, when there is a disruption—and as a tool to mark cutting events in the production of knowledge, where the knowledge process is understood as the construction of a research apparatus, through which the world undergoes changing interventions, and to emphasize the destabilization of the disembodied and isolated subject position of the scientists. Posthuman thinking transcends network theories, and it places the subject in relation to the object and the apparatus in an agential and material semiotically entangled environment (Ernst, ibid).

The Syrian refugee Ashraf Albesh had lost one leg by a bombing, i.e. by a hard, disruptive event. Now in Cologne, having received a prostheses similar to the ones of Oskar Pistorius, he discovered his entanglement with the artificial leg as a means for dancing, an activity he never had done before. He is performing now on stage fathoming the technolimits of a new potential of movements, of agency of his new enhanced body. So an obstruction, a disruption that can induce an unexpected diffractive transformation, is able to change the habitual smooth wavesurface into disturbance giving open space for new possibilities.

Nonetheless, feminist posthumanities reject a transhumanist vision of human enhancement in the direction of immortality: it is a masculinist desire stemming from enlightenment, driven by envy to the birth-giving woman, to realize the disembodied human by use of science, medicine and technology in order to avoid disease, ageing and eventually death. Such a dream of perfection and infinity conflicts with feminist ethics, which takes incompleteness and vulnerability as an integral part of human and non-human existence. Reflecting what is offered by transhumanism, it aims at overcoming the body, its weaknesses and vulnerabilities, and a very one-sided picture of mankind can be stated: talking of surpassing, optimization and obtainable superiority presupposes a competitive society, a masculine-heroic concept of human life. But humanity also includes weakness, vulnerability, incompleteness, inadequacy and limitations. All these imperfections have to be respected, also because life is only meaningful in the context of intricacies, hassle and opposed sadness, the alternation between need and satisfaction, success and failure, between effort and exhaustion.

References

Adorno, T.W.: Theorie der Halbbildung. In: Tiedemann, R. (ed.) Gesammelte Schriften. Band 8: Soziologische Schriften I. Suhrkamp, Frankfurt am Main (1959)

Bacon, F.: New Atlantis. William Raleigh, London (1669)

Barad, K.: Meeting the Universe Halfway. Duke University Press, Durham & London (2007)

Crutzen, C.K.M.: ICT-representations as transformative critical rooms. In: Kreutzner, G., Schelhowe, H. (eds.) Agents of change: virtuality, gender, and the challenge to the traditional university, pp. 87–106. Springer, Opladen (2003)

Crutzen, C.: Der kritisch-transformative Raum "Elitesport" als Phänomen, S. 71–73. In: Marsden, N., Wulf, V., Rode, J., Weibert, A. (eds.) Proceedings of the GEWINN Gender & IT Conference, Heilbronn, pp. 3–10. ACM ICPS publisher, Heilbronn (2018)

Dermühl, K.: The Body Beyond Nature? Exploration, invasive Technologien, gesellschaftliche Implikationen. iF-Schriftenreihe Sozialwissenschaftliche Zukunftsforschung 01/15. Institut Futur, Freie Universität Berlin, Berlin (2015)

Duttweiler, S., Gugutzer, R., Passoth, J.-H., Strübing, J. (eds.): Leben nach Zahlen. Self-Tracking als Optimierungsprojekt? transcript Verlag, Bielefeld (2016). isbn:978-3-8376-3136-4

Ernst, W.: Menschliche und weniger menschliche Verbindungen; Posthumanismus und Gender. FIfF-Kommunikation. **3**, 37–41 (2016)

Friedrich, A.: Outsmarting the transient. In: Gamm, G., et al. (eds.) Jahrbuch Technikphilosophie 2016: List and Death. diaphanes, Zurich (2016)

Groys, B., Vidokle, A. (eds.): Kosmismus. Matthes und Seitz, Berlin (2018)

Harasser, K.: Parahumane Konstellationen von Körper und Technik. FIfF-Kommunikation. **2**, 40–45 (2016)

Haraway, D.: A cyborg manifesto: science, technology, and socialist feminism in the late twentieth century. In: Haraway, D.J. (ed.) Simians, Cyborgs, and Women: The Reinvention of Nature, pp. 149–181. Free Associations Books, London (1991)

Haraway, D.: When Species Meet. University of Minnesota Press, Minneapolis (2007)

Kunzru, H.: You Are Cyborg, Interview with Donna Haraway, Wired. http://www.wired.com/1997/02/ffharaway/ (1997)

Miah, A.: Be very afraid: cyborg athletes, transhuman ideals and posthumanity. J. Evol. Technol. **13**, 20 (2003) http://www.jetpress.org/volume13/miah.html

Nietzsche, F.: Also sprach Zarathustra. München 1999: Deutscher Taschenbuch Verlag/de Gruyter (Kritische Studienausgabe, Bd. 4; Eds. Giorgio Colli und Mazzino Montinari) (1885)

Taureck, B.H.F.: Überwachungsdemokratie: Die NSA als Religion. Verlag Wilhelm Fink, Paderborn (2014)

Chapter 15
Co-creation in Transhuman Realities: Setting the Stage for Transformative Learning

Christian Stary

Abstract This contribution investigates co-creative system design in transhuman settings based on a reflective learning approach. Transhuman settings can be understood as systems of co-creation and co-evolution, laying the groundwork for development processes informed by educational reflection principles. A communication framework and system architecture utilizing explanation facilities of machine learning systems allows for co-constructive intervention and stepwise alignment of objectives, capabilities and interaction, and thereby, induces transformation processes, both for the actors involved and the transhuman setting they are part of. (This chapter is based on the presentation at the IS4SI 2017 Summit Digitalisation for a Sustainable Society in Gothenburg—see also Stary (Multidiscip Digit Publish Inst Proc 1(3):236, 2017).)

1 Introduction

In order to achieve More's vision of transhumanism, intelligent life needs to be evolved beyond its current human form, in particular overcoming human limitations by means of science and technology. Although this transformation should be guided by life-promoting principles and values (More 1990), techno-centric advancements are inherent properties when advocating the improvement of human capacities through advanced technologies.

Bostrom (2009, 2014) envisions self-emergent artificial systems that could finally control the development of intelligence, and thus, human life. Such an approach would extend the human-driven strand of evolving systems with an artificial one. In order to streamline upcoming developments, Kenney and Haraway (2015) have argued for co-construction in settings with humanoid and artificial actors. They

C. Stary (✉)
Business Informatics – Communications Engineering, Johannes Kepler University of Linz, Linz, Austria
e-mail: Christian.Stary@JKU.at

W. Hofkirchner, H.-J. Kreowski (eds.), *Transhumanism: The Proper Guide to a Posthuman Condition or a Dangerous Idea?*, Cognitive Technologies,
https://doi.org/10.1007/978-3-030-56546-6_15

both qualify as species being qualified to act and thus call for "*sym-poiesis rather than auto-poiesis*" (p. 260). Thereby, they challenge system design taking into account humans and technological systems of equal capabilities—transhuman entities need to be considered as dedicated parties in co-creative processes.

As artificial systems may show self-emergent behaviour, interaction processes recognizing this feature need to be considered relevant for system design, triggering co-creation and co-evolving system development. In this chapter, a perspective is taken that looks at underlying principles to guide co-creation involving entities with different ontological grounds on this process. It is based on interactions in the so-called double loop, as proposed by Argyris for organizational learning (Argyris 2000). Major principles stem from andragogy that enable structuring interaction for co-creation in transhuman settings. The application of these principles to transhuman system development is driven by capabilities of reflection both from human actors and from AI (artificial intelligence) systems. Core is the situation-aware behaviour of actors in transhuman settings, as it lays the groundwork for future system behaviours.

This contribution is structured as follows. First, the characteristics of transhuman systems and their development are presented in a condensed way. Then, beliefs and value system underlying human activity systems are addressed featuring double-loop learning processes. The latter can be guided by agogic principles and serve as bridge to establish reflective practice in transhuman settings. Explanation components of transhuman systems can be triggered for clarification of underlying assumptions, decision-making and mutual alignment. Thereby, interventions on beliefs and value systems can be set. As transformative actions they frame co-construction in transhuman settings, and allow conclusive evidence on mutual learning processes, and at the same time, structure future research activities.

2 Transhumanist System Development

Transhumanist proponents are driving the development of their concepts now for the past two decades, aiming to accelerate the evolution of intelligent life beyond its current human form by means of science and technology. The artefacts developed so far integrate findings from different fields, including biotechnology, robotics, information technology, molecular nanotechnology and artificial general intelligence (cf. Goldblatt 2002). Of central interest are brain functions that are considered for whole-brain emulation (cf. Bostrom 2014, p. 30 ff.). Capabilities needed for that are:

- Scanning, ranging from preprocessing/fixation and physical handling to imaging
- Translation, starting with image processing allowing for scan interpretations, and in this way, laying groundwork for software models of the neural system
- Simulation, addressing storage, bandwidth, CPU, (virtual) body and (virtual) environment simulation

As can be concluded from the published transhumanist's declaration (http://humanityplus.org/philosophy/transhumanist-declaration/), the proponents suggest an ethical use of technologies when developing posthuman artefacts.

> 'Sooner or later, the most glaring implementational inefficiencies will have been optimized away, the most promising algorithmic variations will have been tested, and the easiest opportunities for organizational innovation will have been exploited.' (Bostrom 2014, p. 69). Transferring an intellect from a biological brain to a computer system through uploading will thus come to a point in time 'when the rate of technological development becomes so rapid that the progress-curve becomes nearly vertical. Within a very brief time (months, days, or even just hours), the world might be transformed almost beyond recognition. This hypothetical point is referred to as the singularity. The most likely cause of a singularity would be the creation of some form of rapidly self-enhancing greater-than-human intelligence'(http://humanityplus.org/philosophy/transhumanist-faq/ accessed 20.5.2017) (cf. Kurzweil 2006).

The development path to that point in time is considered a sequential 4-step procedure (Bostrom 2014, p. 95f), starting with a *pre-criticality phase*, culminating in "seed AI" being able to improve its own intelligence, independently of human programmers. The subsequent *recursive self-improvement phase* tops human programming skill w.r.t. AI design. "*An intelligence explosion results—a rapid cascade of recursive self-improvement cycles causing AI's capabilities to soar*" (Bostrom 2014, p. 96). It lays the groundwork for the *covert preparation phase* where self-improving AI penetrates information and communication systems through self-propagation. In the final *overt implementation phase* control is organized and set up by self-improving AI systems which establishes singularity.

Besides describing the technology path, transhumanist protagonists envision overcoming human ageing and widening cognitive capabilities. Whole-brain emulation should enable creating substrate-independent minds. Such developments, however, should be guided by risk management and social processes "*where people can constructively discuss what should be done, and a social order where responsible decisions can be implemented*" (see declaration item 4 in the transhumanist declaration). Increasing personal choices over how individuals design their lives based on assistive and complementary technologies (termed "*human modification and enhancement technologies*" in item 9 of the Transhumanist Declaration) is a vision that seems to attract many people. For instance, the Singularity Network (https://www.facebook.com/groups/techsingularity/), one of the largest of hundreds of transhumanist-themed groups on the web, showed an increase from 2011 to 2017 from around 400 to 26.372 in 2017 (as indicated 20.5.2017 on the website).

Transhumanists envision as design entity "*post-humans*" through continuous growth of intelligence that can be *uploaded* to computer systems. When laying the groundwork for levels of consciousness that human brains cannot access so far, posthumans could either be completely synthetic artificial intelligences or composed of many smaller systems augmenting biological human capabilities, which finally cumulates in profound enrichments of human capabilities (cf. More and Vita-More 2013).

Besides applying advanced nano-, neuronal and genetic engineering, artificial intelligence and advanced information management play a crucial role when developing intermediary forms between the human and the posthuman (which are termed *transhumans*). Subjects of design and later on designers themselves due to their self-replicating capability are therefore intelligent machines in an ever-increasing range of tasks, and level of autonomy. Once the continuous replacing of human intelligence by machine intelligence creates machine intelligence superior to single human cognitive intelligence, social integration and balancing control becomes crucial (Bostrom 2014, p. 132). Effective and efficient planning requires social forces for self-emerging phenomena. According to Bostrom, they could either penalize or reward AI developments.

Once humans are involved in this process, this distinct dual perspective could be part of a wider understanding, as, e.g., indicated by Haraway (2015) for human groundings. Humans are rooted in their natural context which in particular influences the evolution of species: "*I am a compost-ist, not a posthuman-ist, we are all compost not posthuman*" (p. 160) Taking into account such categorical binding challenges design approaches due to the intended decoupling of human and machine intelligence. Being aware of the current development and increasing interest in transhumanist developments a learning cycle needs to be triggered in the course of design, that informs development on both an operational and meta-operational level (cf. Argyris 2000).

3 Grounding Transhumanist System Development on Agogic Principles

In learning settings taking into account two adjacent levels, the operational level concerns reflecting concrete development activities (also termed part of single-loop learning—see the next section), e.g. the application of cyber-physical development routines. The second layer, i.e. the meta-operational level, concerns underlying assumptions and value systems that direct and trigger the operational level (and are reflected as part of double-loop learning—see the next section). In particular, the meta-operational level takes into account underlying assumptions and encoded beliefs that guide behaviour of actors in a system, such as the rationale of applying routines.

At that meta-operational level, values or drivers of co-creation and collaboration can be negotiated, as they influence the operational level. It is this level that needs to be addressed for organizing the interaction of transhumans and thus structuring transhuman settings for co-construction. Of fundamental current interest are learning processes of developers as they finally need to be embodied in transhuman developments for creating compost-ists (rather than posthumanists). We first introduce the guiding principles before providing a knowledge development framework and interaction architecture adjusting learning levels in Sect. 4.

If we consider development issues, such as transhuman system design, meaningful to human life and the existence of being human, guidance needs to be provided for all stakeholder groups, defining it as pedagogic, andragogic and gerontagogic issue. Consequently, it poses an agogic challenge to researchers when looking for a scientific approach that can provide a mode of thought that might result in man's overcoming present dilemmas. The Greek idea of the agein, from which the adjective agogic is derived, refers to being together in the world ("mitwelt", "mitsein"). When reference is made to transhuman system development as an agogic approach, the respective guiding principles can be investigated with respect to their applicability.

Starting point for an agogic model can be Dewey's distinction between impulsive, routine and reflective action (cf. Dewey 1910a, b, 1933). Impulsive action is based on trial and error; routine action is based largely on authority and tradition; reflective action is based on "*the active, persistent and careful consideration of any belief or supposed form of knowledge in the light of the grounds that support it*" (Dewey 1933, p. 9). He explains reflective thinking as a "chain" not only involving "*a sequence of ideas but a con-sequence*" of thoughts (ibid., p. 4). In his understanding, acting in open-mindedness and responsibility are consequences of reflective thinking.

Schön (1983) also refers to professional practice. His reflective practitioner approach aims at professional capabilities to handle complex and unpredictable problems of actual practice with confidence, skill and care. According to Adler (1990), a professional practitioner can not only think while acting, but also deal with conflicts involved in situations. Unique or surprising situations are handled through reframing and finding new solutions ("*reflection-in-action*"). This process is (i) a conscious one, though not necessarily articulated in words; (ii) a critiquing one, as it leads to questions and re-structuring; and (iii) immediately significant for action (most important) (cf. Schön 1987, p. 29).

An agogic and situation-aware mindset asserts that an actor's time perspective changes from postponed application of experiences and knowledge to immediacy of application and, accordingly, orientation to acting shifts from subject-centred activities to focused interaction in co-creative settings (cf. Bronfenbrenner 1981). There, several agogic principles apply:

- Activities are set in accordance with the needs of participating actors under the given conditions and capabilities to act.
- Each actor has certain resources that are not only the starting point, but rather design entities. They are accepted to be limited.
- Actors determine their way and pace of developments, as development needs to be balanced with the current conditions. Both active participation and retreat are part of development processes.

It is the latter principle that is of crucial importance for triggering development in co-creative settings. Agogic actors need to embody (cf. Rogers 2003), and thus self-manage

- Empathy as sensitive understanding of others
- Appreciation of another personality without preconditioning acceptance and respect
- Congruence meaning the authenticity and coherence of one's person and behaviour.

Congruence is decisive in making visible system values and their attributes to others. Authenticity refers to meet a person "as a person", to the equal of a person, experiencing a situation with the entire spectrum of channels (perceived impulses, feelings, impression, etc.). Coherence includes judging in how far or at what point in time values or elements can be shared with others, i.e. becoming visible to others. An essential part of congruence is that all participating actors or systems have the same transparent understanding of the co-creative system, including preset conditions and irreversible process design, e.g. normative or role-specific behaviour.

Agogic behaviour in a certain situation, e.g. co-creating transhuman systems, indicates sensing to be crucial for setting cognitive acts intentionally. It means capturing the rationale of acting in terms of perceiving a situation and cognitively reflecting perceived information. Both serve as some preprocessor to acting and are guided by intention and planned action. According to that model, various subsystems are involved in preparatory actions through reflecting on system information and making action from system processing visible to others in the shared setting, e.g. a transhuman system development space.

- WHAT IS IT? What did you see, hear, smell, tasted, feel? What happened, when and how? Can you describe in detail?
- WHAT SHOULD IT BE? Which perspective, which sense do you see? What needs to be achieved? Which priorities do you want to set? What do you want exactly? And why? Which state satisfies you?
- WHY? Which meaning do the observations have for you? Which relations do you recognize? What do you reckon? How can you explain that? What are your conclusions?
- HOW? How to proceed? Which means shall be used? Which tactics shall we chose? What is to be done? Who does what, with what, whom, when and how?

These items stem from Arbeitsagogik.ch and will guide the interaction architecture and framework for transformative learning (cf. Mezirow 2000; Stary 2017) presented in the next section. Of particular interest is the rationale, i.e. the WHY as it refers to underlying assumptions and embedded value systems that become part of knowledge claims when being challenged.

4 Informing Co-creation Through Reflection

In this section, we develop the conceptual foundation of co-creation in transhuman settings. The first subsection introduces the double loop in learning as required for reflection on action. The second subsection approaches reflection processes according to the original concept of the reflective practitioner, in order to structure transformative learning processes. The final subsection introduces the framework and architecture enabling the implementation of mutual learning processes, based on explanation capabilities of transhuman machine learning components.

4.1 Reflection as Trigger of Transformative Learning Cycles

Learning on the operational and meta-operation level has been investigated by Firestone and McElroy (2003) when proposing their iterative and collaborative learning model, the Knowledge Life Cycle (KLC). It intertwines single (operational) and double-loop learning (meta-operational) processes. In case of mismatches on the operational level, e.g. when transhuman systems evolve or are (further) developed in a misleading way, problem detection and formulation induce knowledge processing through problem and knowledge claim formulation (codified beliefs, guiding principles and meta-cognitive elements). By iteratively acquiring and processing information along individual system as well as collective learning processes, informed decision-making on codified knowledge claims is prepared. These claims may survive based on the decision taken and be further processed for diffusion on the collective level, e.g. inducing a certain system behaviour in transhuman settings.

Repositories, such as the distributed organizational knowledge base (DOKB) of the KLC, play a crucial role, both once knowledge w.r.t. knowledge claim evaluation is documented and when phasing accepted knowledge claims into operational procedures, finally changing system behaviour (i.e. the HOW) in transhuman settings. As integrative living design memory, repositories allow reconfiguring previously produced knowledge claims and tie them to running codification schemes, e.g. representing values, and operational (transhuman system development) processes. When stepping beyond operation into a knowledge processing environment on transhuman system development, a double-loop learning process is triggered in this way. It transforms the actors and their relations, i.e. the setting they are operating in.

The KLC bridges the gap between single- and double-loop learning which can be argued to be essential for transhuman system development processes: It allows reflecting on ongoing development processes as well as fundamental design decisions, e.g. deciding whether the point in time for handing over control to machine intelligence has come. In the double-loop meta-information controlling the operational development environment is processed, as knowledge claims are formulated, evaluated, processed and prepared for integrating modifications into running operational procedures. There, further optimization in terms of single-loop learning, e.g. reducing behaviour risks of components, can be performed.

Hence, in case the development of a system cannot satisfy collective needs by means of the development environment, then double-loop learning is initiated. In a cycle, feedback and achievements are exchanged based on triggers from the operational development and leading to informed adjustments or changes (cf. Argyris and Schön 1996). It is the KLC's knowledge processing phases, namely knowledge production and knowledge integration, where several agogic principles are of importance, since underlying drivers of developing transhuman systems need to be addressed explicitly and in a collaborative way. Of particular importance is a system's or actor's balancing of development needs with current conditions, in order to set activities in accordance with the needs and capabilities of participating actors or systems (see the previous section).

Achieving these properties throughout development and evolvement supports congruence in co-creation, as transparency and intelligibility can be improved through meta-layer information or interventions which could in turn affect the collaboration of actors. Once actors of transhuman systems acquire contextual knowledge of operational procedures, they can co-operate in development in an informed way.

4.2 Skilful Reflective Practice

Reflective practice involving meta-operation requires knowledge on how to trigger and implement respective activities. In the following some findings referring to that level are detailed.

Reflective practice starts with the self of actors and propagates formulated knowledge claims to the collective for evaluation. When doing reflective practice is considered as a means of self-efficacy with respect to becoming aware of oneself in a contextual form, re-framing experienced situations and behaviour patterns. It is "*learning through and from experience towards gaining new insights of self and practice*" according to Finlay (2008, p. 1), and needs to be understood as an active, dynamic and action-oriented set of skilfully iterated experience—reflect—conceptualize—apply cycles, which characterize double-loop learning. While linking one experience to the next, it allows substantial change of behaviour rather than opportunistic behaviour adaptation. It sustains while the latter fades away (Fig. 15.1).

By skilful reflective practice an actor becomes a transformative learner, as a practitioner is implicitly involved in "*changing the situation from what it is to something he likes better*" according to Schön (1983, p. 147). Since overall "*the idea of reflective practice leads to a vision of professionals as agents of society's reflective conversation with its situation*" (ibid., p. 353), actors or systems can profit from this technique when interfacing novel situations, such as dealing with machine intelligence embodied in digital artefacts.

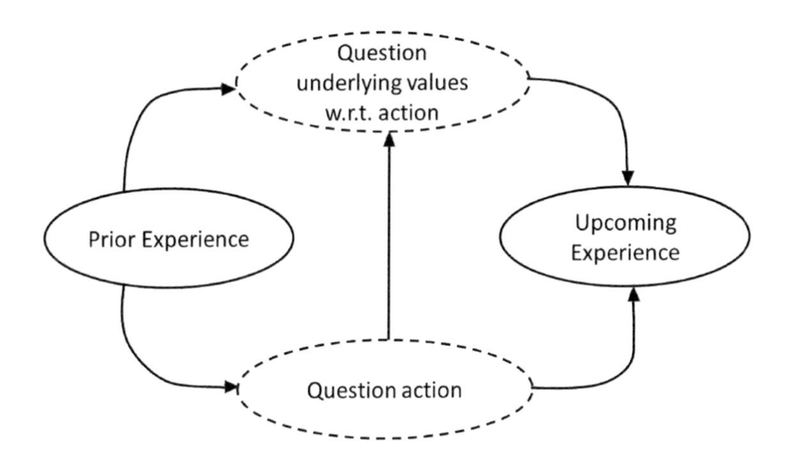

Fig. 15.1 Schematic perspective on reflective practice

Reflective practice as learning tool helps collecting data on experiences, creating coherent explanations or theories, and decomposing and recomposing (complex) systems, until a situation makes sense, and thus provides meaning from experiences. This holds in particular when interacting with digital artefacts, as they have encoded values in their behaviour, which accelerates the need for some kind of inner dialogue or discourse on one's underlying assumptions and values in sociotechnical and transhuman settings. Reflective practice supports (re-)capturing specific views and relating them to evidence individual systems or actors have collected from other situations.

It was John Dewey (1910b) referring to the dual nature of perceived information. When suggesting to study the nature or origin of any belief or "*supposed form of knowledge*", he established reflective practices as a form of questioning the ground that supports belief or supposed form of knowledge. He claimed the WHY questioning is essential for both the situation as it is and as it could be envisioned in future experience. Typical drivers for asking WHY have been identified:

- The quest for emotional regulation and as a consequence the ability to inspire, influence and motivate others
- Demonstrating decision-making capabilities in terms of risk taking and/or systemic impact
- Generating innovation through asking open questions
- To be compassionate to self and others and inspire trust through demonstrating trustworthiness
- Establishing working and working relationships while being challenged by a volatile, uncertain, complex and ambiguous world

Reflecting throughout practice means striving not for just looking back on past actions and events, but taking a conscious look at emotions, experiences, actions and responses. Utilizing that information, the individual knowledge base of an actor or system is enriched, and it can reach a higher level of understanding. Reflective practice is about looking into individual activations, thoughts and emotions, while looking out at the situation experienced (cf. Johns 1994).

The key driver of reflective practice is the objective and making meaning of a perceived situation. In this way, it follows Gestalt therapy, as persons should become fully aware of the driving forces to be and live experiment (cf. Nevis 2017). Thereby, the self of a person plays a crucial role, as Gestalt therapy deals with mechanisms of self-organization, when a person designs itself as part of a living universe. As such, it needs to take responsibility for experiences through which the environment and various contexts of a person's acting are embodied into its life. For interacting with transhumans, in particular, the "*social*" context (successively enabled by technology) needs to be reflected upon.

Bolton (2010) proposed the WHAT? as starting point of reflective practice. When answering this question intermediaries could play a role by revealing a post- or transhuman has been or still is part of a situation. The WHAT requires a description and some kind of representation that can be processed further on. It is followed by SO WHAT?, and finally NOW WHAT?, when closing the loop to WHAT for the upcoming experience. This triad is intended for persons reflecting on ways in which they can personally improve and influence the consequences of their response to the experience.

Like Bolton's three phases most approaches for operationalizing reflective practice require to start focusing on retrospection: "*retrospection: i.e. thinking back about a situation or experience; self-evaluation, i.e. critically analysing and evaluating the actions and feelings associated with the experience, using theoretical perspectives; reorientation, i.e. using the results of self-evaluation to influence future approaches to similar situations or experiences*" according to Quinn (1988, p. 82). The skill of critically reflecting on experience has been termed by Schön (1983) the capacity to reflect on action, namely "*so as to engage in a process of continuous learning*". In line with Kolb's (1984) experiential learning concept, learning allows transforming information into knowledge. This initial phase of reflective practice is composed of

- Re-inhabitation (reliving the experience)
- Reflection through noticing what was going on
- Reviewing it through critically analysing the situation
- Reframing through capturing new understanding

The subsequent stage of reflective practice is reflection in action. It requires recognizing patterns of thoughts, feelings and physical responses as they happen, and using this information to choose what to do moment by moment. Consequently, the ultimate stage is reflection for action, which combines insight with intention to apply learning in reflexive life. Typical triggers or starting events that have been identified for reflective practice (cf. Gibbs 1988) are:

- A disorienting dilemma, such as unexpected job offers or change of relationship, or moving to a new culture
- Self-examination with feelings of fear, anger, guilt or shame

- A critical assessment of assumptions
- Recognition that one's discontent and the process of transformation are shared
- Exploration of options for new roles, relationships and actions
- Planning a certain course of action
- Acquiring knowledge and (new) skills for implementing a plan
- Provisional testing of new roles
- Building competence and self-confidence in new roles and relationships
- Taking a new perspective impacting one's activities

Since reflective practice is a conscious, self-managed learning process, it cannot be established as a direct result of the experience. It requires intentional steps taken by individuals, and thus, personal responsibility for learning.

Gibbs, when following Kolb's experiential learning cycle, has developed a set of well-cited questions, in particular for intertwining theory and practice for mutual enrichment in the context of retro- and introspection. The questions are listed as given in the Reflective Learning Cycle (cf. Kolb 1984):

- DESCRIPTION: When and where did the situation take place? Why were you there? Who else was there? What happened? What did you do? What did not you do? What did they do? What did not they do? What happened?
- FEELINGS: What were your feelings before this situation? What were your feelings at the time? What were your feelings afterwards? What do you think the others in the situation were feeling?
- EVALUATION: What went well/was positive? Why? What did not go so well/ was less positive? Why? What was your contribution? What was the contribution of others?
- ANALYSIS: What assumptions are you making? What insights are now available to you?
- CONCLUSION and ACTION PLAN: What will you do differently? What skills do you need to develop to achieve this? Who and what will support your development in this area?

The cycle consists of 4 stages and comprises dedicated activities (denoted by bullet points):

Stage 1 (Concrete experience). A stakeholder experiences something new for the first time in a certain context. The experience should be an active one, used to test new ideas and procedures.

- Naming the initial experience provides an identification of the reflection.

Stage 2 (Observation of the concrete experience, then reflecting on the experience). Here practitioners should consider the strengths of the experience and areas of development. Practitioners need to form an understanding of what impact their action had on the environment, in both directions—helping and hindering.

- Descriptions capture what happened without making judgements or drawing conclusions.
- Feelings refer to (re)actions without analysing them.
- Evaluation is focusing on what was good or bad about the experience and leads to value judgements.

Stage 3 (The formation of abstract concepts.) The practitioner needs to make sense of what has happened. It should be done through making links between what action had been set, what knowledge the practitioner has and what needs to be learnt. The practitioner should draw on ideas from existing findings to help support development and understanding and draw on support from others and their previous knowledge. Practitioners should modify their ideas or devise new approaches, based on what they have learnt from their observations and wider research.

- Analysis is composed of (i) trying to make sense of the experienced situation, bringing in ideas from outside the experience to help, (ii) identifying what was really going on and (iii) finding out whether different people's experiences were similar or different in important ways.

Stage 4 (The practitioner considers how to put what has been learnt into practice). The practitioner's abstract concepts are made concrete by testing ideas in future situations, resulting in new experiences. The ideas from the observations and conceptualizations are made into active experimentation as they are implemented into future (professional) activities.

- Conclusions are of general and specific nature. General ones aim towards principles derived from the experiences and the analyses that have been made, while specific ones are unique or personal to the situation or the way of acting.
- Personal action plans contain upcoming events in terms of doing something differently in this type of situation the next time, and the steps that one is going to take on the basis of what has been learnt.

The cycle is then repeated on this new action. It is through reflection a practitioner will utilize a repertoire of understandings, images and actions to reframe a (troubling) situation so that problem-solving or alternative actions are generated.

Meaningful learning is facilitated through self-examination of assumptions, patterns of interactions and the operating premises of action. This emphasis on critical reflection can lead to transformational learning exhibited through reflective action. Hence, operational situations and interactions could be viewed as continuous co-creations like "*a dance-like pattern, simultaneously involved in design and in playing various roles in virtual and real worlds, while at the same time remaining detached enough to observe and feel the action that is occurring, and to respond*", according to Tremmel (1993, p. 436).

Consequently, reflective practice is triggered by a WHY, and leads to addressing fundamental assumptions (i.e. the value dimension) aside from the cognitive and emotional dimension of reflection (cf. Thompson and Thompson 2018). They are additionally influenced by the social distance of reflection processes, as they may concern the self or dyadic and group relations (ibid.). Consequently, in transhuman settings these dimensions need to be recognized for implementation for co-creation, even though reflective practice is a concept deeply rooted in humans.

4.3 Utilizing Explanations for Transformative Learning Cycles

In this section, we introduce an approach on how to achieve transformative learning contributing to transhuman system design based on reflective practice. Bostrom (2014) has recognized several strategically relevant tasks and corresponding skill sets when dealing with transhuman intelligence which enables increasingly autonomous ecosystems (cf. Shakir and Aijaz 2017). In the context of this work, i.e. dealing with reflection for co-creation in transhuman settings, of particular interest are the following ones (Bostrom 2014, p. 94):

- *Intelligence amplification*—"*system can bootstrap its intelligence*".
- *Strategizing*: It contains strategic planning, prediction and analysing options for "*optimizing chances of achieving distant goal*".
- *Social manipulation*: It is based on models of social and psychological mechanisms for manipulation and rhetoric persuasion, not only when "*leveraging external resources by recruiting human support*", but also to push others "*to adopt some course of action*".

Bootstrapping system intelligence is one of the core mechanisms of DARPA's explainable AI (XAI) programs as presented to the public in the notes https://www.darpa.mil/attachments/XAIProgramUpdate.pdf, available at the website https://www.darpa.mil/program/explainable-artificial-intelligence). According to these sources, machine learning applications in autonomous systems should be able to explain their decisions and actions to human users, as they will perceive, learn, decide and act on their own (cf. Gunning 2017).

Explainable machine learning aims to understand, appropriately trust and effectively manage an emerging generation of artificially intelligent machine partners. Therefore, machine learning techniques should provide models intelligible to humans, while maintaining a high level of learning performance (prediction accuracy). Ideally, machine learning systems have the ability to

- Explain the rationale of their behaviour
- Document strengths and weaknesses
- Convey how they will behave in the future

Developers try to provide these capabilities by modelling the encoded knowledge. The resulting models are combined with state-of-the-art human–computer interface techniques capable of translating models into understandable and useful

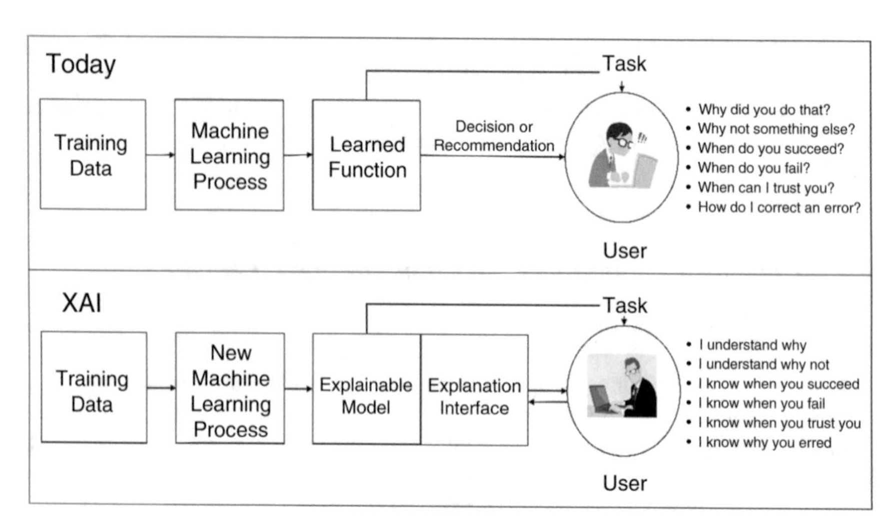

Fig. 15.2 Explainable AI concept according to Gunning (2017)

explanation dialogues for the end user (see Fig. 15.2). The questions human users could be interested are listed on the right top of the figure.

XAI leads to machine learning systems that include some representation of context and the environment in which they operate. Their explanatory models address the capability of machine learning algorithms (i) to classify events of interest in heterogeneous multimedia data and (ii) to construct decision policies for an autonomous system to perform a variety of simulated missions. These areas form the core of reinforcement learning to analyse intelligence.

Doshi-Velez et al. (2017), besides others (cf. Miller et al. 2017), have questioned the accountability of the information delivered by machine learning systems. Although explanations w.r.t. classification and accumulated knowledge are technically feasible in a timely way for explicitly encoded knowledge, their provision needs to be enriched with checks for validity and correctness.

Given the differences between human and machine learning processes, Doshi-Velez et al. (2017) identify situations which require more human activity w.r.t. explanation, and others in which machine learning systems have a higher standard of explanation. Their study distinguishes extrinsic from intrinsic factors to the decision-maker.

Extrinsic factors concern the significance of a decision, and thus the extent to which an explanation will inform future action. They challenge human and AI decision-makers the same way, whereas intrinsic factors may vary significantly between humans and machine learning systems (see Table 15.1). Doshi-Velez et al. (2017) consider handling intrinsic factors crucial, since preplanning explanations makes the difference. Humans will, in the course of making a decision, generate and store the information needed to explain that decision later if doing so becomes useful.

Table 15.1 Human and AI systems' strengths and weaknesses w.r.t. explanability, according to Doshi-Velez et al. (2017)

Capabilities for explanation	Humans	Machine learning systems
Strengths	Ability to provide explanation post hoc	Behaviour is reproducible Operate without social pressure
Weaknesses	May be inaccurate May be not reliable May act under social pressure	Requires up-front engineering Requires encoded knowledge including explicit taxonomies

Machine learning systems traditionally do not automatically store information about their decision-making process, in order to optimize data storage and to protect privacy. In order to generate explanations, resources have to be allocated to their generation, starting with system design. However, in the course of design developers have to take care that the inmates are not running the asylum (cf. Miller et al. 2017). Based on empirical data when eliciting knowledge from legal scholars, computer scientists and cognitive scientists in collective settings, Doshi-Velez et al. (2017) found consistency about when explanations are required, while there is a fair amount of consistency in what the abstract form of an explanation needs to be. The latter they consider particularly helpful when creating machine learning systems providing explanations:

> In the colloquial sense, any clarifying information can be an explanation. Thus, we can "explain" how an AI makes decision in the same sense that we can explain how gravity works or explain how to bake a cake: by laying out the rules the system follows without reference to any specific decision (or falling object, or cake). When we talk about an explanation for a decision, though, we generally mean the reasons or justifications for that particular outcome, rather than a description of the decision-making process in general. In this paper, when we use the term explanation, we shall mean a human-interpretable description of the process by which a decision-maker took a particular set of inputs and reached a particular conclusion. ...
>
> In addition to this formal definition of an explanation, an explanation must also have the correct type of content in order for it to be useful. As a governing principle for the content an explanation should contain, we offer the following: an explanation should permit an observer to determine the extent to which a particular input was determinative or influential on the output. Another way of formulating this principle is to say that an explanation should be able to answer at least one of the following questions: What were the main factors in a decision? This is likely the most common understanding of an explanation for a decision. In many cases, society has prescribed a list of factors that must or must not be taken into account in a particular decision. For example, we many want to confirm that a child's interests were taken into account in a custody determination, or that race was not taken into account in a criminal prosecution. A list of the factors that went into a decision, ideally ordered by significance, helps us regulate the use of particularly sensitive information. (Doshi-Velez et al. 2017, p. 2f)

As decisions depend on the (type of) information available, it should be part of the discourse and documented in a repository available for all parties involved. By looking at the effect of changing that information on the output and comparing it to the previous results, it can be inferred whether a knowledge claim can be accepted.

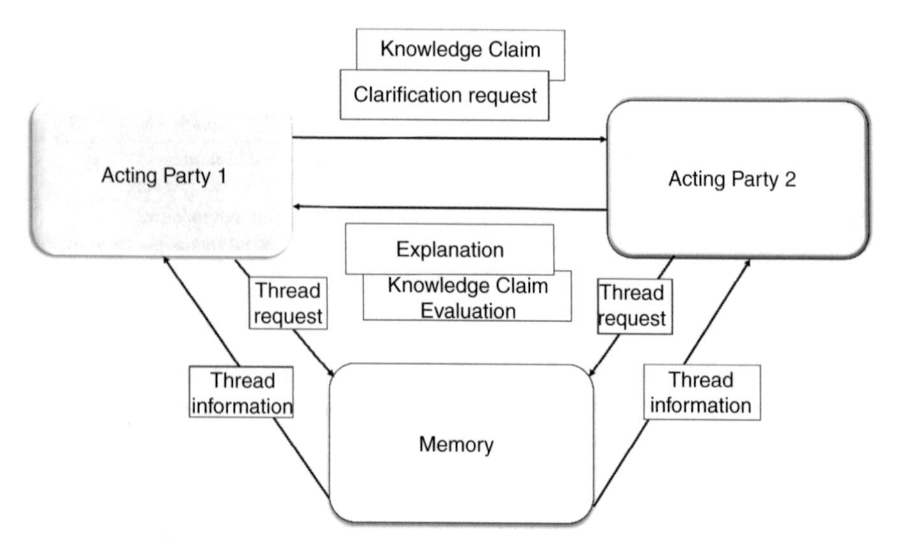

Fig. 15.3 Architecture for transformation learning in transhuman settings

Furthermore, when making decision-relevant information transparent, it can be identified whether a specific factor is influential and, finally, is determinative for specific actions.

Figure 15.3 shows the abstract architecture for transformative learning interaction based on explanation capabilities of actors and reflective practice. Acting Party 1 and 2 are two transhuman system components. Acting Party 1 could either formulate a knowledge claim or a request for clarification referring to some behaviour of Acting Party 2 or result from its decision-making. It receives an explanation or the result of an individual knowledge claim evaluation. This pattern reoccurs until the next action is set by Acting Party 1 in a reflected way.

In the memory component of the transhuman setting, all relevant information when handling requests is kept, including situational context, decision-making principles and sent and received messages. Various threads of interactions according to the learning cycles and reflections are preserved to trace back behaviour sequences and situational experiences in transhuman settings. Such a memory function avoids re-producing decisions and knowledge claim evaluations and enables situation-sensitive comparisons over time.

At the level of abstraction, the architecture for transformative setting shows the utilization of explanations without revealing the internal processing scheme of the acting transhuman parties. The architecture is open for different patterns of internal behaviour of actors. In addition, it follows the KLC as well as the XAI structure separating explanation-relevant information from the actual evaluation and decision-making processes. In particular, explaining machine learning is quite different to intelligent behaviour of transhuman components. The latter produces some predictions that should match the transhuman setting. In contrast, creating an

Table 15.2 Major elements to be processed for clarification and claim evaluation

Learning stage	Reflective practice (see Sect. 3)	Agogic aspect (see Sect. 2)
Labelling	Identification of reflection	What is?
Description	Capturing the situation	What is? How?
Feelings	Capturing status at the time of experience	What is?
Evaluation	Explicating rationale(s) for experienced activities	Why? How?
Analysis	Explicating assumptions and insights from the experienced situation	Why? How?
Conclusive action plan	Preparing further activities including identifying crucial context factors	What should be?

explanation requires a human-interpretable set of steps and rules that allows tracing how an input to the transhuman component is processed to generate the resulted prediction.

As shown in Table 15.2, transformative learning follows the stages of skilful reflective practice (starting with reflection on action) and addresses agogic aspects. The latter have also been considered relevant for reflective practice, e.g. Now What refers to What Should It Be (as detailed in Sects. 2 and 3). The stages listed in the table correspond to process steps of the acting parties in Fig. 15.3 and the information that is processed and exchanged between them to set the stage for co-creation.

Considering the rectangles in Fig. 15.3 as behaviour abstractions of transhuman system elements (cf. Fleischmann et al. 2012), once they are modelled and validated, they can be automatically executed by subject-oriented workflow engines or actor technologies, such as UeberFlow (cf. Krenn and Stary 2016). When executing the models, system behaviour can follow the experience—reflect—conceptualize—apply learning cycle (as proposed by Kolb) and thus establish reflective practice and double-loop learning as inherent features of transhuman settings.

5 Conclusions

Co-creative system design in transhuman settings can be based on a reflective learning approach. Such settings are driven by actors exchanging requests and information on encoded knowledge for decision-making. It serves as baseline reference when clarifying or justifying behaviour. In this contribution, the underlying concepts have been detailed and embodied in system development informed by learning principles. In terms of co-creation, the knowledge production and evaluation of knowledge claims can be structured for transforming transhuman system actors. The architecture lays the groundwork for implementing major principles of conversational knowledge creation in future research, as proposed by Wagner (2004, p. 270):

- *Open*: In case development content is found to be incomplete or poorly organized, any actor or system should edit it the way it fits individually.
- *Incremental*: Development content can be linked to other development content, enforcing system thinking and contextual inquiry.
- *Organic*: The structure and content of a system under development is open to continuous evolution.
- *Mundane*: A certain number of conventions and features need to be agreed for shared access to development content.
- *Universal*: The mechanisms of further development and organizing are the same as creating so that any actor, system or system developer can be in both roles, an operating and a development system.
- *Overt*: The output suggests the input required to reproduce development content.
- *Unified*: Labels are drawn from a flat space so that no additional context is required to interpret them.
- *Precise*: Development content items are titled with sufficient precision to avoid most label or name clashes.
- *Tolerant*: Interpretable behaviour is preferred to error messages.
- *Observable*: Activities involving development content or structure, specifications, can be watched and reviewed by other stakeholders, both on the cognitive and social level.
- *Convergent*: Duplication can be discouraged or removed by identifying and linking similar or related development content.

This list helps ensuring the coherence of transformative learning activities when co-creation occurs in arbitrary composed networks of transhuman systems or actors. The properties are to be operationalized in further research activities, based on a platform supporting the execution of transformative learning activities.

References

Adler, S. The reflective practitioner and the curriculum of teacher education. In: Proceedings Annual Meeting of the Association of Teacher Educators, Las Vegas, 29 pages, available at eric.ed.gov. http://files.eric.ed.gov/fulltext/ED319693.pdf (1990). Accessed 20 May 2017

Argyris, C.: Double-loop learning. In: Wiley Encyclopedia of Management, pp. 115–125. Wiley, Hoboken, NJ (2000)

Argyris, C., Schön, D.A.: Organizational Learning II: Theory, Method and Practice. Addison-Wesley, Reading, MA (1996)

Bolton, G.: Reflective Practice: Writing and Professional Development. Sage, London (2010)

Bostrom, N.: The future of humanity. In: Olsen, J.-K.B., Selinger, E., Riis, S. (eds.) New Waves in Philosophy of Technology, pp. 186–216. Palgrave Macmillan, New York (2009)

Bostrom, N.: Superintelligence: Paths, Dangers, Strategies. Oxford University Press, Oxford (2014)

Bronfenbrenner, U.: Die Ökologie der menschlichen Entwicklung. Klett-Cotta, Stuttgart (1981)

Dewey, J.: Educational Essays. Cedric Chivers, Bath (1910a)

Dewey, J.: How We Think. D.C. Heath, Boston (1910b)

Dewey, J.: How We Think: A Restatement of the Relation of Reflective Thinking to the Educative Process. Heath & Co, Boston (1933)

Doshi-Velez F, Kortz M, Budish R, Bavitz C, Gershman S, O'Brien D, Schieber S, Waldo J, Weinberger D, Wood A.: Accountability of AI under the Law: the Role of Explanation. Berkman Klein Center Working Group on Explanation and the Law, Berkman Klein Center for Internet & Society working paper. http://nrs.harvard.edu/urn-3:HUL.InstRepos:dash.current.terms-of-use#LAA (2017)

Finlay, L.: Reflecting on reflective practice, PBPL CETL, Open University [Online]. http://www.open.ac.uk/opencetl/resources/pbpl-resources/finlay-l-2008-reflecting-reflective-practice-pbpl-paper-52 (2008). Accessed 16 July 2018

Firestone, J.M., McElroy, M.W.: Key Issues in the New Knowledge Management. Routledge, New York (2003)

Fleischmann, A., Schmidt, W., Stary, C., Obermeier, S., Börger, E.: Subject-Oriented Business Process Management. Springer, Cham (2012)

Gibbs, G.: Learning by Doing: A Guide to Teaching and Learning Methods. Further Education Unit, London (1988)

Goldblatt, M.: DARPA's programs in enhancing human performance. In: Roco, M.C., Schummer, J. (eds.) Converging Technologies for Improving Human Performance: Nanotechnology, Biotechnology, Information Technology and the Cognitive Science, pp. 337–341. National Science Foundation, NBIC-report, Arlington, VA (2002)

Gunning, D.: Explainable Artificial Intelligence (XAI). Defense Advanced Research Projects Agency (DARPA), Arlington, VA (2017)

Haraway, D.: Anthropocene, capitalocene, plantationocene, chthulucene: making kin. Environ. Hum. **6**(1), 159–165 (2015)

Johns, C.: A philosophical basis for nursing practice. In: Johns, C. (ed.) The Burford NDU Model: Caring in Practice. Blackwell Scientific Publications, Oxford (1994)

Kenney, M., Haraway, D.: Anthropocene, capitalocene, plantationocene, chthulucene: Donna Haraway in conversation with Martha Kenney. In: Davis, H., Turpin, E. (eds.) Art in the Anthropocene: Encounters Among Aesthetics, Environments and Epistemologies, pp. 229–244. Open Humanity Press, London (2015)

Kolb, D.: Experiential Learning as the Science of Learning and Development. Prentice Hall, Upper Saddle River, NJ (1984)

Krenn, F., Stary, C.: Exploring the potential of dynamic perspective taking on business processes. Complex Syst. Inf. Model. Q. (8), 15–27 (2016)

Kurzweil, R.: The Singularity Is Near: When Humans Transcend Biology. Viking, New York (2006)

Mezirow, J.: Learning to think like an adult: core concepts of transformation theory. In: Mezirow, J. (ed.) Learning as Transformation, pp. 3–34. Jossey-Bass, San Francisco (2000)

Miller, T., Howe, P., Sonenberg, I.: Explainable AI: beware of inmates running the asylum. In: Proceedings IJCAI-17 Workshop on Explainable AI (XAI), vol. 36, pp. 36–42 (2017)

More, M.: Transhumanism: towards a futurist philosophy. Extropy. **6**, 6–12 (1990)

More, M., Vita-More, N.: The Transhumanist Reader: Classical and Contemporary Essays on the Science, Technology, and Philosophy of the Human Future. Wiley, Hoboken, NJ (2013)

Nevis, E.C.: Gestalt Therapy: Perspectives and Applications. CRC Press, Boca Raton, FL (2017)

Quinn, F.M.: Reflection and reflective practice. In: Davies, C., Finlay, L., Bullman, A. (eds.) Changing Practice in Health and Social Care. Sage, London (1988/2000)

Rogers, C.R.: Die klientenzentrierte Psychotheraphie. Kindler, München (2003)

Schön, D.A.: The Reflective Practitioner: How Professionals Think in Action, vol. 5126. Basic Books, New York (1983)

Schön, D.A.: Educating the Reflective Practitioner. Jossey-Bass, San Francisco (1987)

Shakir, M. Z., Aijaz, A.: IoT, robotics and blockchain: towards the rise of a human independent ecosystem. http://cn.committees.comsoc.org/files/2015/11/Newsletter_May2017.pdf (2017)

Stary, C.: Agogic principles in trans-human settings. Multidiscip. Digit. Publish. Inst. Proc. **1**(3), 236 (2017)

Thompson, S., Thompson, N.: The Critically Reflective Practitioner. Macmillan International Higher Education, London (2018)

Tremmel, R.: Zen and the art of reflective practice in teacher education. Harv. Educ. Rev. **63**(4), 434–459 (1993)

Wagner, C.: Wiki: a technology for conversational knowledge management and group collaboration. Commun. Assoc. Inf. Syst. **13**(1), 265–289 (2004)

Printed in the United States
by Baker & Taylor Publisher Services